AF890688

Lectures on
STOCHASTIC CONTROL AND NONLINEAR FILTERING

By

M. H. A. DAVIS

Lectures delivered at the
INDIAN INSTITUTE OF SCIENCE, BANGALORE
under the
T.I.F.R. — I.I.Sc. PROGRAMME IN APPLICATIONS OF MATHEMATICS

Notes by

K. M. RAMACHANDRAN

Published for the
TATA INSTITUTE OF FUNDAMENTAL RESEARCH

SPRINGER-VERLAG
Berlin Heidelberg New York Tokyo
1984

Author

M. H. A. DAVIS
Department of Electrical Engineering
Imperial College of Science and Technology
London SW 7
UNITED KINGDOM

© TATA INSTITUTE OF FUNDAMENTAL RESEARCH, 1984

ISBN 3-540-13343-7 Springer-Verlag, Berlin. Heidelberg. New York. Tokyo
ISBN 0-387-13343-7 Springer-Verlag, New York. Heidelberg. Berlin. Tokyo

No part of this book may be reproduced in any form by print, microfilm or any other means without written permission from the Tata Institute of Fundamental Research, Colaba, Bombay 400 005

Printed by M. N. Joshi at The Book Centre Limited, Sion East, Bombay 400 022 and published by H. Goetze, Springer-Verlag, Heidelberg, West Germany

PRINTED IN INDIA

PREFACE

These notes comprise the contents of lectures I gave at the T.I.F.R. Centre in Bangalore in April/May 1983. There are actually two separate series of lectures, on controlled stochastic jump processes and nonlinear filtering respectively, and the corresponding two parts of these notes are almost disjoint. They are united however, by the common philosophy (if that is not too grand a word for it) of treating Markov processes by methods of stochastic calculus, and I hope the reader will, at least, be convinced of the usefulness of this and of the 'extended generator' concept in doing calculations with Markov processes.

The first part is aimed at developing optimal control theory for a class of Markov processes called piecewise-deterministic (PD)processes. These were only isolated rather recently but seen general enough to include as special cases practically all the non-diffusion continuous time processes of applied probability. Optimal control for PD processes occupies a curious position just half way between deterministic and Stochastic optimal control theory in such a way that no standard theory from either side is adequate to deal with it. The only applicable theory that exists at all is very recent work of D. Vermes based on the generalized dynamic programming ideas of R.B. Vinter and R.M. Lewis, and this is what I have attempted to describe here. Undoubtedly, further development of control theory for PD processes will be a fruitful field of enquiry.

Part II concentrates on the "pathwise" theory of filtering for diffusion processes and on more sophisticated extensions of it due primarily to H. Kunita. The intriguing point here is to see how stochastic partial differential equations can be dealt with by stochastic flow theory through what amounts to a "doubly stochastic" version of the Feynman-Kac formula. Using this, Kunita has given an elegant argument to show the existence of smooth conditional densities under Hörmander-type conditions. This is included. Ultimately, it rests on results obtained by Bismut and others using Malliavin calculus, since one needs a version of the Hörmander theorem which is valid for continuous (rather than C^{∞}) t - dependence of the coefficients. It was unfortunately impossible to go into such questions in the time available.

I would like to thank Professor K.G. Ramanathan for his kind invitation to visit Bangalore and K.M. Ramachandran for his heroic efforts at keeping up-to-date notes on a rapidly accumulating number of lectures, and for preparing the final version of the present text. I would also like to thank the students and staff of the T.I.F.R. Centre and of the I.I.Sc. Guest House for their friendly hospitality which made my visit such a pleasant one.

CONTENTS

PREFACE

PART I. STOCHASTIC JUMP PROCESSES AND APPLICATIONS

CHAPTER I. STOCHASTIC JUMP PROCESSES

§0. Introduction 1

§1. Martingale Theory for Jump Processes 2

§2. Some Discontinuous Markov Processes 24

CHAPTER II. OPTIMAL CONTROL OF PD PROCESSES .. 47

PART II. FILTERING THEORY

§0. Introduction 66

§1. Linear and Nonlinear Filtering Equations 67

§2. Pathwise Solutions of Differential Equations 84

§3. Pathwise Solutions of the Filtering Equations 90

BIBLIOGRAPHY 106

PART I STOCHASTIC JUMP PROCESSES AND APPLICATIONS

CHAPTER I

STOCHASTIC JUMP PROCESSES

§0. INTRODUCTION

Stochastic jump processes are processes with piecewise constant paths. The Poisson process, the processes arising in Inventory problems (stocks of items in a store with random ordering and replacement) and queueing systems (arrivals at a queue with each customer having random demand for service) are examples of stochastic jump processes. Our aim here is to develop a theory suitable for studying optimal control of such processes.

In Section 1, martingale theory and stochastic calculus for jump processes are developed. Gnedenko-Kovalenko [16] introduced piecewise-linear process. As an example of such a process, consider virtual waiting time process (VWT) for queueing systems, where VWT(t) is the time customer arriving at time t would have to wait for service, see Fig. (1).

Later Davis [7] and Vermes [25] introduced the concept of piecewise deterministic processes which follow smooth curves (not necessarily straight lines) between jumps. In Section 2, we will study some applications to piecewise-deterministic processes. The idea there is to derive Markov properties, Dynkin's formula, infinitesimal generators etc., using the calculus developed in Section 1.

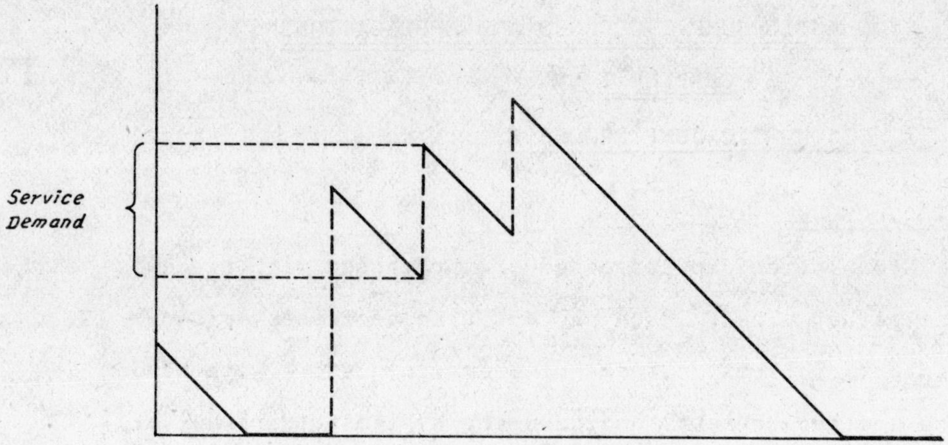

Fig. (1) Arrival time of customers

1. MARTINGALE THEORY FOR JUMP PROCESSES

Let $(X, \mathcal{S})$ be a Borel space.

DEFINITION 1: A jump process is defined by sequences $T_1, T_2, T_3, \ldots, Z_1, Z_2, Z_3, \ldots$ of random variables, $T_i \in \mathbb{R}_+$ and $T_{i+1} > T_i$ a.s. and $Z_i \in (X, \mathcal{S})$. Set

$$T_\infty = \lim_{k \to \infty} T_k.$$

Let z_o, z_∞ be fixed elements of X. Define the path $(x_t)_{t \geq o}$ by

$$x_t = \begin{cases} z_o & \text{if } t < T_1 \\ Z_i & \text{if } t \in [T_i, T_{i+1}[\\ z_\infty & \text{if } t \geq T_\infty. \end{cases}$$

Then the probability structure on the process is determined by either joint distribution for $(T_i, Z_i, i = 1,2,\ldots)$ or specifying

(i) distribution of (Z_1, T_1)

(ii) for each $k = 1,2,3,\ldots,$ conditional distribution of $(S_k, Z_k | T_{k-i}, i = 1,2,\ldots)$, where $S_k = T_k - T_{k-i}$ is the kth inter-arrival time.

We will start studying the process (x_t) having a single jump, i.e.,

$$x_t = \begin{cases} z_o & \text{if } t < T(\omega) \\ Z(\omega) & \text{if } t \geq T(\omega). \end{cases}$$

If $T = \infty$, let $Z = z_\infty$, a fixed point of X.

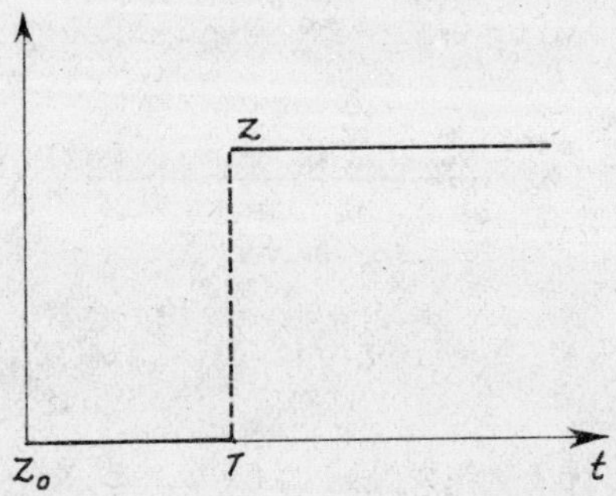

Fig.(2)

Define the probability space (Ω, F, P) as the canonical space for T, Z, i.e.,

$$((\mathbb{R}_+ \times X) \cup \{(\infty, z_\infty)\}, \mathcal{B}(\mathbb{R}_+) * S, \{(\infty, z_\infty)\}, \mu)$$

where μ is a probability measure on

$$((\mathbb{R}_+ \times X) \cup \{(\infty, z_\infty)\}, \mathcal{B}(\mathbb{R}_+) * S, \{(\infty, z_\infty)\}).$$

The random function (x_t) generates the increasing family of σ - fields (F_t^o), i.e.,

$$F_t^o = \sigma\{x_s, s \leq t\}.$$

We suppose

$$\mu(([o, \infty] \times \{z_o\}) \cup \{o\} \times X) = 0.$$

This assumption guarantees that the process x_t does jump at its jump time T, i.e.,

$$P(T > o \text{ and } Z \neq z_o) = 1.$$

Recall that an $\mathbb{R}_+$ - valued random variable τ is a stopping time of a filtration F_t, if $(\tau \leq t) \in F_t$, $\forall$ t. Let

$$F_t = \text{Completion of } F_t^o \text{ with all } F_\infty^o \text{ - null sets.}$$

<u>PROPOSITION 1.1</u>: T <u>is not an</u> F_t^o <u>stopping time, but</u> T <u>is an</u> F_t <u>stopping time</u>.

<u>PROOF</u> Let $A = \{Z = z_o\}$ and K be any set in X. Then

$$x_s^{-1}(K) = \begin{cases} ((s,\infty] \times X) \cup ([o,s] \times \{z_o\}) & \text{if } z_o \in K \\ & \text{and } Z(E - A) \cap K = \phi \\ ([s,\infty] \times X) \cup ([o,s] \times K) & \text{if } z_o \in K \\ & \text{and } Z(E - A) \cap K \neq \phi. \\ [o,s] \times K & \text{if } z_o \notin K. \end{cases}$$

where $E = \mathbb{R}_+ \times X - A$.

Clearly $[o, t] \times X$ cannot be in the σ - algebra generated by sets of the above form. So T is not an F_t^o stopping time. Let $B = X - \{z_o\}$. By assumption, $P(A) = 0$; so $A \in F_t$.

$$x_t^{-1}(B) = [o,t]\times X - A \in F_t^o.$$

So

$$[o,\ t]\times X \in F_t.$$

But $\{T \le t\} = [o,\ t]\times X$. Hence T is an F_t stopping time.

#

It can be seen that

$$F_t = B[o,t] * S\ U(]t,\infty]\times X)\ U \text{ null sets of } F_\infty^o.$$

The stopped σ - field F_T is given by

$$F_T = \{G \in F_\infty : G \cap (T \le t) \in F_t,\ \forall\ t\}.$$

Clearly

$$F_T = F_\infty.$$

<u>DEFINITION 1.2</u>: A process (M_t) is an F_t - martingale if $E|M_t| < \infty$ and for $s \le t$

$$E[M_t | F_s] = M_s \quad \text{a.s.}$$

(M_t) is a local F_t - martingale if there exists a sequence of stopping times $S_n \uparrow \infty$ a.s. such that $M_t^n := M_{t \wedge S_n}$ is a uniformly integrable martingale for each n; here $t \wedge S_n := \text{Min}(t,\ S_n)$.

<u>PROPOSITION 1.2</u>: <u>If</u> M_t <u>is a local martingale and</u> S <u>is a stopping time such that</u> $S \ge T$ a.s., <u>then</u> $M_S = M_T$ a.s.

<u>PROOF</u>: Let S_n be stopping times such that $S_n \uparrow \infty$ a.s. Then $M_{t \wedge S_n}$ is u.i. martingale. Let $M_t^n = M_{t \wedge S_n}, \forall\ n$. Then by optional sampling theorem

$$E[M_S^n\ F_{T \wedge S_n}] = M_T^n ;$$

but

$$F_T = F_\infty.$$

So

$$E[M_S^n | F_T] = M_S^n.$$

Also

$$\lim_{n \to \infty} M_T^n = M_T$$

and

$$\lim_{n \to \infty} M_S^n = M_S.$$

So

$$M_S = M_T \quad \text{a.s.}$$

PROPOSITION 1.3: Suppose τ is an F_t - stopping time. Then there exists $t_o \in \mathbb{R}_+$ such that $\tau \wedge T = t_o \wedge T$ a.s.

PROOF: If τ is a stopping time, then $(\tau \wedge T \leq t) \in F_t$, $\forall\, t$. But if $T \wedge \tau$ is not constant a.s. on $(\tau \leq T)$, then

$$(\tau \wedge T \leq t) \cap (]t,\infty] \times X) \subsetneq]t,\infty[\times X \quad \text{for some } t \in \mathbb{R}_+.$$

But $]t,\infty] \times X$ is an atom of F_t. This contradicts the fact that τ is a stopping time. So

$$\tau \wedge T = t_o \wedge T \quad \text{a.s.}$$

The general definition of a stopped σ - field is that if U is a stopping time. Then

$$F_U = \{A \in F | A \cap (U \leq t) \in F_t, \ \forall\, t\}.$$

But this is an implicit definition of the σ - field.

EXERCISE 1.1: Suppose $\tau = t_o \wedge T$. Show that

(i) $F_\tau = F_{t_o}$

(ii) $F_\tau = \sigma\{X_{\tau \wedge s}, s \geq o\}$.

DEFINITION 1.3: For $A \varepsilon S$, define

$$F^A(t) = \mu(]t,\infty] \times A)$$

and

$$F(t) = F^X(t) = P[T > t].$$

Note that $F(o) = 1$ and $F(.)$ is monotone decreasing and right continuous. Define

$$c = \begin{cases} \inf\{t: F(t) = 0\} \\ +\infty \text{ if } \{t: F(t) = 0\} = \phi. \end{cases}$$

PROPOSITION 1.4: Suppose $(M_t)_{t \geq o}$ is an F_t local martingale. Then

(a) if $c = \infty$ or, $c < \infty$ and $F(c-) = 0$, then M_t is a martingale on $[o,c[$.

(b) if $c < \infty$, $F(c-) > o$, then (M_t) is a uniformly integrable martingale. Here $F(c-) = \lim_{t \uparrow c} F(t)$.

PROOF: (a) If $\tau_k \geq T$ a.s. for some k, then

$$M_{t \wedge \tau_k} = M_{t \wedge \tau_k \wedge T} = M_{t \wedge T} = M_t.$$

So M_t is a u.i. martingale. Hence suppose $P[\tau_k < T] > o$ for all $k(*)$; then by Proposition 1.3,

$$\tau_k \wedge T = t_k \wedge T \text{ for some fixed } t_k$$

and $t_k < c$ because of (*). Also $t_k \uparrow c$ since $\tau_k \uparrow \infty$. So

$$M_{t \wedge \tau_k} = M_{t \wedge \tau_k \wedge T} = M_{t \wedge T \wedge t_k} = M_{t \wedge t_k}.$$

Hence $M_{t \wedge t_k}$ is a u.i. martingale. So $(M_t)_{t < c}$ is a martingale.

(b) $c < \infty$, $F(c_-) > o$.

$$F(c_-) = P(T = c)$$

$P(T = c) > o$; so it must be the case that $t_k = c$ for some k. Otherwise $P(\tau_k < c) \geq F(c-) > o$; so "$\tau_k \uparrow \infty$ a.s." fails.

For this k,

$$M_{t \wedge t_k} = M_t.$$

So (M_t) is a u.i. martingale.

#

Our main objective is to show that all local martingales can be represented in the form of "stochastic integrals". So we introduce some "elementary martingales" associated with the process (x_t). For $A \in S$ and $t \in \mathbb{R}_+$, define

$$p(t,A) = I_{(t \geq T)} I_{(Z \in A)}$$

$$\tilde{p}(t,A) = -\int_{]o, T \wedge t[} \frac{1}{F(s_-)} dF^A(s).$$

<u>PROPOSITION 1.5</u>: <u>Let</u> $q(t,A) = p(t,A) - \tilde{p}(t,A)$. <u>Then</u> $(q(t,A))_{t \geq o}$ <u>is an</u> F_t - <u>martingale</u>, i.e., $\tilde{p}(t,A)$ <u>is the "compensator" of the point process</u> $p(t,A)$.

PROOF: (Direct calculation). Take $t > s$, then

$$E[p(t, A) - p(s,A)|F_s] = I_{(s < T)} \frac{F^A(s) - F^A(t)}{F(s)} .$$

$$E[\tilde{p}(t,A) - \tilde{p}(s,A)|F_s] = + I_{(s < T)}\{\frac{F(t)}{F(s)} \int_{]s,t]} \frac{dF^A(u)}{F(u-)}$$

$$- \frac{1}{F(s)} \int_{]s,t]} \int_{]s,r]} \frac{dF^A(u)}{F(u-)} dF(r)\}$$

and

$$\int_{]s,t]} \int_{]s,r]} \frac{dF^A(u)}{F(u-)} dF(r) = \int_{]s,t]} \frac{1}{F(u-)} \int_{[u,t]} dF(r)dF^A(u)$$

$$= \int_{]s,t]} \frac{1}{F(u-)} (F(t) - F(u-))dF^A(u)$$

$$= F(t) \int_{]s,t]} \frac{dF^A(u)}{F(u-)} + F^A(t) - F^A(s).$$

So

$$E[q(t, A) - q(s,A)|F_s] = 0.$$

#

Another expression for $\tilde{p}(t,A)$: We have $F^A(.) << F(.)$ (i.e. $F^A(.)$ is absolutely continuous w.r.t. $F(.)$). So there exists a function $\lambda(s,A)$ such that

$$F^A(o) - F^A(t) = - \int_{]o,t]} \lambda(s,A)\, dF(s).$$

In fact

$$\lambda(s,A) = P(Z \in A|T = s).$$

Suppose X is such that a regular version of this conditional probability exists (which is the case, since X is Borel space). Then $\frac{-dF^A(s)}{F(s-)} = \lambda(s,A)d\Lambda(s)$ where $d\Lambda(s) = \frac{-dF(s)}{F(s-)}$. Then

$$\tilde{p}(t,A) = \int_{]o,T \wedge t]} \lambda(s,A)\, d\Lambda(s).$$

Stochastic Integrals

Let $\mathcal{I}$ denote the set of measurable functions $g: \Omega \to \mathbb{R}$ such that $g(\infty, z_\infty) = 0$.

(a) Integrals w.r.t. $p(t,A)$: Suppose N_t is a counting process. Since its sample functions are monotone increasing and there is a one-to-one correspondence between monotone increasing functions and measures, and since in this case, mass is concentrated at the jump points and they are only countable; the function N_t defines a random measure on $(\mathbb{R}, \mathcal{B}(\mathbb{R}))$ say, $\pi = \sum_i \delta_{T_i}$ where δ_x is the Dirac measure at x. Similarly, the one jump process can be identified with the random measure $\delta_{(T,x_T)}$ on $R_+ \times X$. So we can define Stieltjes integrals of the form $\int g(t,x)\, p(dt,dx)$ for suitable integrands $g \in \mathcal{I}$ as

$$\int_\Omega g(t,x)\, p(dt,dx) = g(T,x_T).$$

We say $g \in L_1(p)$ if

$$E \int_\Omega |g(t,x)|\, p(dt,dx) < \infty$$

and denote

$$||g||_{L_1(p)} = E \int_\Omega |g(t,x)|\, p(dt,dx)$$

Clearly $g \in L_1(p)$ if and only if

$$\int_{\mathbb{R}_+ \times X} |g(t,x) \;\; d\mu < \infty$$

(b) Integrals w.r.t. $\tilde{p}(t,A)$:

Recall $\tilde{p}(t,A) = \int_{]o,T \wedge t]} \lambda(s,A)d\Lambda(s)$.

So we define

$$\int_\Omega g(t,x)\tilde{p}(dt,dx) = \int_{]o,T]} \int_X g(t,x)\lambda(t,dx)d\Lambda(t)$$

and say

$$g \in L_1(\tilde{p}) \quad \text{if} \quad \int_\Omega |g(t,x)|\tilde{p}(dt,dx) < \infty$$

and

$$\| g \|_{L_1(\tilde{p})} = \int_\Omega |g(t,x)|\tilde{p}(dt,\, dx).$$

PROPOSITION 1.6:

$$\| g \|_{L_1(p)} = \| g \|_{L_1(\tilde{p})}$$

and so

$$L_1(p) = L_1(\tilde{p}).$$

PROOF:

$$\| g \|_{L_1(\tilde{p})} = -\int_{\mathbb{R}_+} \int_{]o,T]} \frac{1}{F(s_-)}|g(s,x)|d\mu(s,x)dF(T)$$

$$= \int_\Omega \frac{1}{F(s_-)}|g(s,x)|(-\int_{[s,\infty]} dF(t))d\mu(s,x)$$

$$= \int_\Omega |g(s,x)|d\mu(s,x)$$

$$= \| g \|_{L_1(p)}.$$

#

Define

$$L_1^{loc}(p) = \{g \in I | g(s,x)I_{(s \le t)} \in L_1(p), \forall\ t < c\}$$

$$L_1^{loc}(\tilde{p}) = \{ g \in I \mid g(s,x) I_{(s \leq t)} \in L_1(\tilde{p}), \quad \forall\ t < c \}.$$

Clearly

$$L_1^{loc}(p) = L_1^{loc}(\tilde{p}).$$

Following is the main result of this section, which gives an integral representation for F_t local martingales.

PROPOSITION 1.7: All F_t - local martingales are of the form

$$M_t = \int_\Omega (g(s,x)\ I_{(s \leq t)})\ dq(s,x)$$

$$= \int_\Omega (g(s,x)\ I_{(s \leq t)})\ dp(s,x) - \int_\Omega (g(s,x)\ I_{(s \leq t)})\ d\tilde{p}(s,x).$$

for some $g \in L_1^{loc}(p)$.

We need the following result.

LEMMA 1.1: Suppose $(M_t)_{t > o}$ is a u.i. F_t martingale with $M_o = 0$. Then there exists a function $h: \Omega \to \mathbb{R}$ such that

$$E|h(T,Z)| < \infty \tag{1}$$

and

$$M_t = I_{(t \geq T)} h(T,Z) - I_{(t<T)} \frac{1}{F(t)} \int_{]o,t] \times X} h(s,z) \mu\ (ds,dz) \tag{2}$$

PROOF: If (M_t) is a u.i. martingale, then $M_t = E[\xi | F_t]$ for some F_∞ - measurable r.v. ξ and from the definition of F_∞, we have

$$\xi = h(T,Z) \quad \text{a.s.}$$

for some measurable $h: \Omega \to \mathbb{R}$. Expression (1) is satisfied since M_t is u.i., and $M_o = 0$ implies

$$\int_\Omega h.d\mu = 0 . \tag{3}$$

Now

$$M_t = E[h(T, Z)|F_t]$$

$$= I_{(t \geq T)} h(T, Z) + I_{(t < T)} \frac{1}{F(t)} \int_{]t,\infty] \times X} h(s,x) \, \mu(ds, dx). \tag{4}$$

From (3) and (4), we have (2).

#

For $g \in L_1^{loc}(p)$, the stochastic integral

$$M_t^g = \int_{]o,t] \times X} g(s,x) \, q(ds, dx)$$

is defined by

$$M_t^g = \int_{\mathbb{R}^+ \times X} I_{(s \leq t)} g(s,x) p(ds,dx) - \int_{\mathbb{R}^+ \times X} I_{(s \leq t)} g(s,x) \, \tilde{p}(ds, dx).$$

Then the question is whether M_t given by (2) is equal to M_t^g for some g. As a motivation to the answer consider the following example.

<u>EXAMPLE 1.1</u>: Let $(X, S) = (\mathbb{R}, \mathcal{B}(\mathbb{R}))$ and

$$\mu(ds, dx) = \psi(s,x) \, ds \, dx.$$

Then

$$M_t^g = I_{(t \geq T)} \{g(T,Z) - \int_o^T \int_{\mathbb{R}} \frac{1}{F(s)} g(s,x) \psi(s,x) dx \, ds\}$$

$$- I_{(t < T)} \{\int_o^t \int_{\mathbb{R}} \frac{1}{F(s)} g(s,x) \psi(s,x) dx \, ds\} \tag{5}$$

If M_y^g given by (5) is equal to M_t given by (2), then the coefficients of $I_{(t \geq T)}$ and $I_{(t < T)}$ must agree. Comparing the coefficients of $I_{(t \geq T)}$, we require

$$h(t,z) = g(t,z) - \int_o^t \int_{\mathbb{R}} \frac{1}{F(s)} g(s,x)\psi(s,x)\, dx\, ds.$$

Let

$$\eta(t) = h(t,z) - g(t,z).$$

Define

$$\gamma(t) = \int_{\mathbb{R}} \psi\, h\, dx$$

and

$$f(t) = \int_{\mathbb{R}} \psi\, dx.$$

Then

$$\eta(t) = \int_o^t \frac{1}{F(s)} \left(\int_{\mathbb{R}} \mathbf{h}(s,z)+\eta(s))\psi(s,x)dx\right) ds$$

$$= \int_o^t \frac{1}{F(s)} \gamma(s)\, ds + \int_o^t \frac{1}{F(s)} \eta(s)f(s)ds\ ;$$

that is

$$\frac{d}{dt}\eta(t) = \frac{f(t)}{F(t)}\eta(t) + \frac{1}{F(t)}\gamma(t)$$

$$\eta(o) = 0$$

which has a unique solution

$$\eta(t) = \int_o^t \Phi(t,s)\frac{1}{F(s)}\gamma(s)\, ds,$$

where

$$\Phi(t,s) = \exp \int_s^t \frac{f(u)}{F(u)}\, du$$

$$= \frac{F(s)}{F(t)}, \quad \text{since} \quad f(t) = -\frac{dF(t)}{dt}\ .$$

So

$$\eta(t) = \frac{1}{F(t)} \int_o^t \gamma(s)\, ds.$$

Hence

$$g(t,z) = h(t,z) + \frac{1}{F(t)} \int_o^t \int_{\mathbb{R}} h(s,x)\psi(s,x)dx\, ds \qquad (6)$$

Now it can be checked that with this choice of g the coefficients of $I_{(t < T)}$ in (2) coincides with that of (5). So $M_t = M_t^g$.

#

Now we can prove the general case given in Proposition 1.7.

<u>PROOF OF PROPOSITION 1.7</u>:

<u>Case 1</u>: $c < \infty$, $F(c\_) > o$. Take a local martingale M_t with $M_o = 0$. Then (M_t) is u.i. So $M_t = E[h(T,Z)|F_t]$ for some measurable h such that $E|h| < \infty$, $E\, h = 0$. Then we claim that $M_t = M_t^g$ where

$$g(t,z) = h(t,z) + I_{(t < c)} \frac{1}{F(s)} \int_{]o,t] \times X} h(s,z)\mu(ds,dz).$$

But this can be seen algebraically following similar calculations as of the example 1.1. Now to show that $g \in L_1^{loc}(p)$.

$$\begin{aligned}\int|g|d\mu &\leq \int|h|d\mu - \int_{]o,c[} \frac{1}{F(t)} \int_{]o,t] \times X} |h|d\mu\, dF(t) \\ &\leq \int|h|d\mu - \frac{1}{F(c\_)} \int_{]o,c[} \int_{]o,t] \times X} |h|d\mu\, dF(t) \\ &\leq \int|h|d\mu + \frac{1}{F(c\_)} \int_{]o,c[\times X} (F(t)-F(c\_))|h|d\mu \\ &\leq (1 + \frac{1}{F(c\_)}) \int|h|\, d\mu < \infty.\end{aligned}$$

Case 2: $c = \infty$, or $c < \infty$ and $F(c_-) = 0$. Then from proposition 1.4, M_t is a martingale on $[o,c[$, and so it is u.i. on $[o,t]$ for $t < c$. Therefore

$$M_s = E[h(T,Z)|F_\infty]$$

for some function h satisfying

$$\int_{]o,t] \times X} |g(s,x)|d\mu(s,x) < \infty \quad \text{for all} \quad t < c.$$

Define $g(s,x)$ as in (6). Then calculations as in case 1. Show that $M_s = M_s^g$ for $s \leq t < c$. Now

$$\int_{]o,t] \times X} |g|d\mu \leq \int_{]o,t] \times X} |h|d\mu - \int_{]o,t]} \frac{1}{F(s)} \int_{]o,s] \times X} |h|d\mu \, dF(s)$$

$$\leq \int_{]o,t] \times X} |h|d\mu (1 - \int_{]o,t]} \frac{1}{F(s)} dF(s))$$

$$< \infty \quad \text{for} \quad t < c.$$

Hence $g \in L_1^{loc}(p)$.

Conversely, suppose $g \in L_1^{loc}(p)$. Then it can be checked that M_t^g is a local martingale.

<u>REMARK 1.1</u>: If $g \in L_1(p)$ then M_t^g is a martingale. But the result does not say M_t is a martingale if and only if $M = M^g$ for $g \in L_1(p)$, it only characterizes <u>local</u> martingales.

<u>REMARK 1.2</u>: All preceding results hold if z_o is a random variable; then μ should be taken as conditional distribution of (T, Z) given z_o.

<u>The multi-jump case</u>: The process x_t has jump times $T_1, T_2, \ldots$ with corresponding states $Z_1, Z_2, \ldots$. Let $(Y, \mathcal{Y})$ denote the measurable space

$$(Y,\mathcal{Y}) = ((\mathbb{R}_+ \times X) \cup \{(\infty, z_\infty)\}, \ \sigma\{\mathcal{B}(\mathbb{R}_+) * S, \{(\infty, z_\infty)\}\}).$$

Define

$$\Omega = \prod_{i=1}^{\infty} Y_i, \quad \Omega_k = \prod_{i=1}^{k} Y_i$$

$$F^o = \sigma\{\prod_{i=1}^{\infty} \mathcal{Y}_i\}$$

where $(Y_i, \mathcal{Y}_i)$ denote a copy of $(Y, \mathcal{Y})$. Let

$$S_k(\omega) = T_k(\omega) - T_{k-1}(\omega)$$

and

$$w_k(\omega) = (S_1(\omega), Z_1(\omega), \ldots S_k(\omega), Z_k(\omega)).$$

Then

$$T_k(\omega) = \sum_{i=1}^{k} S_i(\omega)$$

$$T_\infty(\omega) = \lim_{k \to \infty} T_k(\omega).$$

As before, $(x_t(\omega))_{t \geq o}$ is defined by

$$x_t(\omega) = \begin{cases} z_o & \text{if} \quad t < T(\omega) \\ Z_k & \text{if} \quad t \in [T_k(\omega), T_{k+1}(\omega)[\\ z_\infty & \text{if} \quad t \geq T_\infty(\omega) \end{cases}$$

A probability measure μ on (Ω, F^o) is specified by a family $\mu^i: \Omega_{i-1} \times \mathcal{Y} \to [o,1]$ (with $\Omega_0 = \phi$) satisfying

(i) $\mu^i(.;\Gamma)$ is measurable for each fixed Γ,

(ii) $\mu^i(\omega_{i-1}(\omega);.)$ is a probability measure on $(Y,\mathcal{Y})$ for each fixed $\omega \in \Omega$,

(iii) $\mu^i(w_{i-1}(\omega);(\{o\} \times X) \cup (\mathbb{R}_+ \times \{Z_{i-1}(\omega)\})) = 0$ for all ω,

(iv) $\mu^i(w_{i-1}(\omega); \{(\infty, z_\infty)\}) = 1$ if $S_{i-1}(\omega) = \infty$.

Then for $\Gamma \in Y$ and $\eta \in \Omega_{i-1}$, μ is defined by

$$\mu[(T_1, Z_1) \in \Gamma] = \mu^1(\Gamma)$$

$$\mu[(S_i, Z_i) \in \Gamma | w_{i-1} = \eta] = \mu^i(\eta;\Gamma) \quad i = 2,3...$$

Notice that as in the single jump case, here (iii) ensure that two "jump times" T_{i-1}, T_i do not occur at once and that the process x_t does effectively jump at its jump times (iv) ensures that $\mu[Z_k = z_\infty | T_k = \infty] = 1$.

As before, $F_t^o = \sigma\{x_s, s \leq t\}$ and

F_t = completion of F_t^o with all μ - null sets of F^o.

PROPOSITION 1.8:

(i) $F_\infty = F$, the completion of F^o

(ii) $F_{T_n} = \prod_{i=1}^{n} Y_i \times \prod_{i=n+1}^{\infty} Y_i$.

The idea here is to reduce everything to one jump case. That is, the process "restarts" at each T_k. We need the following result.

PROPOSITION 1.9:

$$F_{(T_{k-1}+t) \wedge T_k} = F_{T_{k-1}} \vee \sigma\{x_{(T_{k-1}+s) \wedge T_k}, 0 \leq s \leq t\}.$$

Proof of this is an application of the "Galmarino test" (Dellacherie and Meyer [12], theorem IV, pp 100).

Recall that in one jump case $F_{t_o \wedge T} = F_{t_o}$. Now we conjecture that, if $U = (T_{k-1} + t_o) \wedge T_k$, then

$$F_U = (\prod_{i=1}^{k-1} Y_i) * Y_{t_o}^k * (\prod_{i=k+1}^{\infty} Y_i),$$

where

$$Y_{t_o}^k = S * B[o, t_o] \cup (X \times]t_o, \infty]).$$

As an example, see the following exercise.

EXERCISE 1.2: Consider a point process with $k = 2$, and take the probability space as $\mathbb{R}_+^2$. Then

$$x_t = \sum_{i=1}^{2} I_{(t \geq T_i)}.$$

Then

(a) Show that

$$F_t = \text{Borel sets in } \{S_1 + S_2 \leq t\}$$
$$+ (A \times \mathbb{R}_+) \cap B + [t,\infty] \times \mathbb{R}_+, \; A \in B(\mathbb{R})$$

where

$$B = \{S_1 + S_2 \geq t\} \cap \{S_2 \leq t\}.$$

(b) With $U = (T_1 + t_o) \wedge T_2$, show that

$$F_U = B(\mathbb{R}_+ \times [o, t])$$
$$+ \{(A \times \mathbb{R}_+) \cap (\mathbb{R}_+ \times]t_o,\infty]) : A \in B(\mathbb{R})\}.$$

ELEMENTARY MARTINGALES, COMPENSATORS

Define

$$p(t,A) = \sum_i I_{(t \geq T_i)} I_{(Z_i \in A)}$$

which counts the jump of (x_t) ending in the set A. Define

$$\phi_1^A(s) = -\int_{]o,s]} \frac{1}{F^1(u)}\, dF^{1A}(u)$$

where

$$F^1 = \mu^1(]t,\infty] \times X)$$

and

$$\phi_k^A(w_{k-1};\ s) = -\int_{]o,s]} \frac{1}{F^k(u_-)}\, dF^{kA}(u)$$

where

$$F^{kA}(u) = \mu^k(w_{k-1};\ ]u,\ \infty] \times A).$$

Now define

$$\tilde{p}(t,A) = \phi_1^A(T_1) + \phi_2^A(w_1;\ S_2) + \ldots\ \phi_j^A(w_{j-1}; t-T_{j-1}(\omega))$$

$$\text{for}\quad t \in\ ]T_{j-1},\ T_j].$$

<u>EXERCISE 1.3</u>: Consider a renewal process

$$x_t = \sum_i I_{(t \geq T_i)}$$

and S_i's are independent, $P(S_i > t) = F(t)$ is continuous. Then show that the compensator for x_t is

$$\tilde{p}(t) = -\ \ell n(F(S_1)F(S_2)\ldots F(S_{k-1})F(t-T_{k-1}))$$

$$\text{for}\quad t \in\ ]T_{k-1},\ T_k]\ ,$$

and

$x_t - \tilde{p}(t)$ is a martingale.

<u>EXAMPLE 1</u>: If $F(t) = e^{-\alpha t}$, then $\tilde{p}(t) = \alpha t$.

PROPOSITION 1.10: For fixed k, and $A \in \mathcal{S}$,

$$q(t \wedge T_k, A) = p(t \wedge T_k, \quad A) - \tilde{p}(t \wedge T_k, \quad A), \ t \geq 0$$

is an $\mathcal{F}_t$ - martingale.

PROOF: Calculations as in proposition 1.5.

The class of integrands $\mathcal{I}$ consists of measurable function $g(t, x, \omega)$ such that

$$g(t,x,\omega) = \begin{cases} g^1(t,x) \quad , & t \leq T_1(\omega) \\ g^k(w_{k-1}, t-T_{k-1}, x), & t \in]T_{k-1}(\omega), T_k(\omega)], \\ \quad 0 \quad , & t \geq T_\infty(\omega) \end{cases}$$

for some function g^k such that $g^1(\infty, x) = g^k(w_k; \infty, x) \equiv 0$. Such g's are $\mathcal{F}_t$ - predictable processes. Now we define $L_1(p)$, $L_1(\tilde{p})$, etc. exactly as in one jump case:

$$\int g \, dp = \sum_i g(T_i, Z_i)$$

$$L_1(p) = \{g \in \mathcal{I}: E \sum_i |g(T_i, Z_i)| < \infty\}$$

$$\int g \, d\tilde{p} = - \sum_k \int_{]o, T_k - T_{k-1}] \times X} g(\omega_{k-1}, s, x) \lambda(\omega_{k-1}, s, dx) \frac{dF^k(s)}{F^k(s_-)}$$

where

$$\lambda(\omega_{k-1}, s, A) = \frac{dF^{kA}}{dF^k}(s)$$

$g \in L_1^{loc}(p)$ if there exists a sequence of stopping times $\sigma_k \uparrow T_\infty$ a.s. and $g \, I_{(t \leq \sigma_n)} \in L_1(p)$, $\forall$ n. For $g \in L_1^{loc}(p)$ we define

$$M_t^g = \int_{]o,t] \times X} g(s,x) \, q(ds, dx)$$

$$= \int_{]o,t] \times X} g(s,x) \, p(ds,dx) - \int_{]o,t] \times X} g(s,x) \tilde{p}(ds,dx).$$

PROPOSITION 1.11: If $g \in L_1^{loc}(p)$ then there exists a sequence of stopping times $\tau_n < T_\infty$ such that $\tau_n \uparrow T_\infty$ and $M^g_{t \wedge \tau_n}$ is a u.i. martingale for each n.

PROOF: Take $\tau_n = n \wedge T_n \wedge \sigma_n$. Then the result follows by direct calculations using **the optional sampling theorem.**

Now let $(M_t)_{t \geq o}$ be a u.i. F_t - martingale. Then

$$M_t = M_{t \wedge T_1} + \sum_{k=2}^{\infty} (M_{t \wedge T_k} - M_{T_{k-1}}) I_{(t \geq T_{k-1})}, \qquad (7)$$

because this is an identity if $t < T_\infty$ and the right-hand side is equal to $\lim M_{T_k}$ if $t \geq T_\infty$. Here we have $M_{T_\infty -} = M_{T_\infty}$. Now we state the main result.

THEOREM 1.1: Let (M_t) be a local martingale of F_t. Then there exists $g \in L_1^{loc}(p)$ such that

$$M_t - M_o = \int_{]o,t] \times X} g(s,x,\omega)\, q(ds,dx).$$

PROOF: Suppose first that M_t in a u.i. martingale.

Define

$$x_t^1 = M_{t \wedge T_1}$$

$$x_t^k = M_{(t + T_{k-1}) \wedge T_k} - M_{T_{k-1}}, \quad k = 2,3,\ldots$$

Then from (7)

$$M_t = \sum_{k=1}^{\infty} x^k_{(t - T_{k-1}) \vee o}.$$

We can now use proposition 1.7 to represent each x^k. Fix k and define for $t \geq o$,

$$H_t = F_{(t + T_{k-1}) \wedge T_k}.$$

Then x_t^k is an H_t martingale. Then there exists a measurable function h^k such that

$$x_t^k = E(h^k(\omega_{k-1}; S_k, Z_k) | H_t).$$

Then using proposition 1.7, there exists $g^k(\omega_{k-1}; s,z)$ such that

$$x_t^k = \int_{]o,t] \times X} g^k(\omega_{k-1}; s,z)\, q^k(ds, dz)$$

where $q^k(t,A) = q((t + T_{k-1}) \wedge T_k, A)$ and $g^k \in L_1^{loc}(p^k)$ for all ω^{k-1} a.s. Piecing these results together for $k = 1,2,3,\ldots.$ gives the desired representation with $g = (g^k)$. It remains to prove that $g \in L_1^{loc}(p)$ as defined, for which we refer to Davis [6].

If (M_t) is only a local martingale with associated stopping time sequence $\tau_n \uparrow \infty$ such that $M_{t \wedge \tau_n}$ is a u.i. martingale, apply the above arguments to $M_{t \wedge \tau_n}$ to complete the proof.

<u>COROLLARY 1.1</u>: <u>If $T_\infty = \infty$ a.s. then the result says (M_t) is a local martingale of F_t if and only if $M_t = M_t^g$ for some</u> $g \in L_1^{loc}(p)$.

<u>REMARK 1.3</u> : It would be useful to determine the exact calss of integrands g required to represent u.i. martingales (as opposed to local martingales) when the jump times T_i are totally inaccessible, Boel, Varaiya and Wong [4] show that $\{M^g, g \in L_1(p)\}$ coincides with the set of u.i. martingales of integrable variation. It seems likely that this coincides

with the set of u.i. martingales if $E\,p(t,E) < \infty$ for all t (a somewhat stronger condition than $T_i \to \infty$ a.s.) but no proof of this is available as yet.

§2. SOME DISCONTINUOUS MARKOV PROCESSES

Extended Generator of a Markov Process

Let the process $x_t \in (E,\mathcal{E})$, some measurable space. Then (x_t,F_t) is a Markov process if for $s \leq t$

$$E[f(x_t)|F_s] = E[f(x_t)|x_s]. \text{ a.s.}$$

A transition function $p(s,x,t,\Gamma)$ is a function such that

$$\begin{aligned} p(s,x_s,t,\Gamma) &= P(x_t \in \Gamma|x_s) \\ &= E[I_\Gamma(x_t)|x_s] \text{ a.s. for } t \geq s. \end{aligned}$$

p satisfies the Chapman-Kolmogorov equation

$$p(s,x,t,\Gamma) = \int_E p(s,x,u,dy)p(u,y,t,\Gamma) \text{ for } s \leq u \leq t.$$

Not every Markov process has a transition function, but usually one wants to start with transition function and construct the corresponding process. This is possible if $(E,\mathcal{E})$ is a Borel space (required to apply Kolmogorov extension theorem; refer Wentzel [27]). One constructs a Markov family,

$$\{P_{x,s}, (x,s) \in E \times \mathbb{R}_+\}, P_{x,s}$$

being the measure for the process starting at $x_s = x$. All measures $P_{s,x}$ have the same transition function p. Denote by $E_{x,s}$ integration w.r.t. $P_{x,s}$. Let $B(E)$ be the set of bounded measurable functions

$$f: E \to \mathbb{R} \text{ with } \|f\| = \sup_{x \in E} |f(x)|.$$

Define

$$T_{s,t} f(x) = E_{x,s}[f(x_t)], \quad s \leq t.$$

$T_{s,t}$ is an operator on $\mathcal{B}(E)$ such that

(i) it is a contraction, $\|T_{s,t} f\| \leq \|f\|$, $T_{s,t}1 = 1$.

(ii) Semi group property: $r \leq s \leq t$,

$$T_{r,t} = T_{r,s} T_{s,t}$$

for

$$\begin{aligned} T_{r,s}(T_{s,t}f)(x) &= E_{x,r}[E_{x_s,s}(f(x_t))] \\ &= E_{x,r}[E(f(x_t)|F_s)] \\ &= E_{x,r}(f(x_t)) \\ &= T_{r,t} f(x). \end{aligned}$$

$T_{s,t}$ is time invariant if $T_{s+r,t+r} = T_{s,t}$ **for** all $r \geq -s$. Then $T_{s,t} = T_{o,t-s} \equiv T_{t-s}$. So T is a one parameter family; this happens when the transition function is time invariant, i.e., $p(s,x,t,\Gamma) = p(s+r, x, t+r,\Gamma)$. Then we get a one parameter family of measures $(P_x, x \in E)$ and the connection is

$$T_t f(x) = E_x f(x_t); \; T_o f = f.$$

Let

$$\mathcal{B}_o(E) = \{f \in \mathcal{B}(E): \|T_t f - f\| \to 0, \; t \downarrow 0\}.$$

An operator $\mathring{A}$ with domain $\mathcal{D}(\mathring{A}) \subset \mathcal{B}_o(E)$ is the <u>strong infinitesimal generator</u> of T_t if

$$\lim_{t \downarrow o} \left\| \frac{1}{t}(T_t f - f) - \mathring{A} f \right\| = o.$$

So

$$\mathring{A} f = \frac{d}{dt} T_t \; f(x)|_{t=o}.$$

Take $f \in \mathcal{D}(\mathring{A})$. Then

$$\lim_{t \downarrow o} \frac{1}{t} (T_t \; T_s f - T_s f) = \lim_{t \downarrow o} \frac{1}{t} (T_{t+s} f - T_s f)$$
$$= \lim_{t \downarrow o} T_s \frac{1}{t} (T_t f - f)$$
$$= T_s \mathring{A} f.$$

So $f \in \mathcal{D}(\mathring{A})$ implies $T_s f \in \mathcal{D}(\mathring{A})$ and $\mathring{A} T_s f = T_s \mathring{A} f$. So we get backward Kolmogorov equation

$$\frac{d}{ds} T_s f = \mathring{A} (T_s f). \tag{1}$$

The main results of the "analytic theory" of Markov semigroups are the following:

(i) Hille-Yosida theorem: Necessary and sufficient conditions for an operator $\mathring{A}$ to be the generator of some semigroup.

(ii) If $\mathring{A}$ satisfies these conditions, then $\mathcal{D}(\mathring{A})$ is dense in $\mathcal{B}_o(E)$ and $(\mathring{A}, \mathcal{D}(\mathring{A}))$ determines T_t (via the so called resolvent operator).

<u>NB</u>. The domain $\mathcal{D}(\mathring{A})$ provides essential information.

Integrating (1), we get <u>Dynkin's formula</u>

$$T_t f(x) - f(x) = \int_o^t T_s \mathring{A} f(x) \, ds$$

i.e.

$$E_x f(x_t) - f(x_o) = E_x \int_o^t \mathring{A} f(x_s) ds, \; f \in \mathcal{D}(\mathring{A}).$$

<u>PROPOSITION 2.1</u>: <u>If</u> $f \in \mathcal{D}(\mathring{A})$ <u>then the process</u>

$$C_t^f = f(x_t) - f(x_o) - \int_o^t \mathring{A} f(x_s) \, ds$$

<u>is a martingale</u>.

PROOF: For $t \geq s$

$$E[C_t^f - C_s^f | F_s] = E[f(x_t) - f(x_s) - \int_s^t \mathring{A} f(x_u) du | F_s]$$

$$= E_{x_s} f(x_t) - f(x_s) - E_{x_s} \int_s^t \mathring{A} f(x_s) ds$$

$$= 0.$$

DEFINITION 2.1: Let $M(E)$ be the set of measurable functions $f: E \to \mathbb{R}$. Then $(A, \mathcal{D}(A))$ with $\mathcal{D}(A) \subset M(E)$ is the extended generator of (x_t) if C_t^f is a local martingale, where

$$C_t^f = f(x_t) - f(x_o) - \int_o^t A f(x_s) ds.$$

This is an extension of $(\mathring{A}, \mathcal{D}(\mathring{A}))$ in that $\mathcal{D}(\mathring{A}) \subset \mathcal{D}(A)$ and $\mathring{A} f = A f$ for $f \in \mathcal{D}(\mathring{A})$. We have uniqueness of A in the following sense. Write

$$f(x_t) = f(x_o) + \int_o^t A f(x_s) ds + C_t^f.$$

This shows that $(f(x_t))$ is a "special semi-martingale" (=local martingale + predictable bounded variation process). The decomposition is unique. So, if B is another generator then

$$\int_o^t (A f(x_s) - B f(x_s)) ds = 0 \quad P_x \quad \text{a.s.} \quad \forall \quad t.$$

Thus $A f(x) = B f(x)$ except on a set of potential zero, where a set Γ has potential zero if

$$E_x \int_o^\infty I_\Gamma (x_s) ds = o \quad \forall \quad x.$$

EXAMPLE 2.1: Suppose $x_t \in \mathbb{R}^d$ satisfies

$$dx_t = b(x_t) dt + \sigma(x_t) dw_t$$

with standard Ito conditions. If $f \in C^2$, then

$$df(x_t) = \Big(\sum_i b_i(x_t)\frac{\partial f}{\partial x_i}(x_t) + \frac{1}{2}\sum_{i,j}(\sigma\sigma')_{ij}\frac{\partial^2 f}{\partial x_i \partial x_j}\Big)\,dt + \int_o^t \nabla f'\sigma\, dw.$$

So $C^2 \subset \mathcal{D}(A)$ and

$$Af(x) = \sum_i b_i(x)\frac{\partial f}{\partial x_i} + \frac{1}{2}\sum_{ij}(\sigma\sigma')_{ij}\frac{\partial^2 f}{\partial x_i \partial x_j}$$

$$C_t^f = \int_o^t \nabla f'\sigma\, dw.$$

NB: This is not a characterisation of $\mathcal{D}(A)$.

REMARK 2.1: If we had required C_t^f to be a martingale rather than a local martingale in definition 2.1, then not every $f \in C^2$ would be in $\mathcal{D}(A)$ because of the properties of Ito integrals.

EXERCISE 2.1: For $i = 1,2,\ldots,$ let N_t^i be a Poisson process with rate λ_i where $\sum_i \lambda_i < \infty$. Define

$$X_t = \sum_{i=1}^{\infty} \ell_i N_t^i$$

where $\ell_i \geq o$ and

$$\sum_{i=1}^{\infty} \ell_i \lambda_i = r < \infty.$$

Find the extended generator of x_t.

(This is also an example where jump times are not isolated).

Piecewise-linear Markov Process

Gnedenko-Kovalenko introduced the concept of piecewise-linear Markow process. Later, Vermes [24] simplified the definition as follows. A piecewise linear process is a two component Markov process $(x_t) = (\nu_t, \xi_t)$ where ν_t is integer-valued and ξ_t takes values in an interval $[a_n, b_n]$ of the real line if $\nu_t = n$ (b_n may be $+\infty$). Let E be the state space, i.e., $E = \{(n,\xi) \in Z \times R: \xi \in [a_n, b_n]\}$. Then the probabilistic description is that if the motion starts at $(n, z) \in E$ and x_t is given by $\nu_t = n$, $\xi_t = z + t$ for $t < T_1$, the first jump time. "Spontaneous jumps" happen at rate $\lambda(x_t)$, i.e., probability of "jump occurs in $(t, t + dt)$" is $\lambda(x_t)\,dt$, and process must jump if $\xi_{t-} = b_n$. Let the transition measure be given by $Q(A; x)$ for $A \in B(E)$. Then x_{T_1} is selected from the probability distribution $Q(A; x_{T_1-})$. After a jump, motion restarts as before. Thus the law of the process is determined by specifying the intervals $[a_n, b_n]$, the jump intensity $\lambda(x)$ and the transition measure $Q(A; x)$.

EXAMPLE 2.2: Non-stationary countable state Markov Process (ξ_t)

(ξ_t) takes integer values with time-dependent transition rates $a_{ij}(t)$ such that

$$P[\xi_{t+h} = i \mid \xi_t = j] = a_{ij}(t)\,\delta + o(\delta),\ i \neq j.$$

Then $x_t = (\xi_t, t)$ is a PL process with no barriers, i.e., $a_n = 0$, $b_n = +\infty$.

EXAMPLE 2.3: Countable state process with non-exponential sojourn times

Here, jump times of the process (x_t) form a renewal process with inter arrival density $b(.)$ and transition matrix $q_{ij} = P[x_{T_k} = i | x_{T_k^-} = j]$. This is a PL process with $\nu_t = x_t$, and ξ_t the time since last arrival. The jump rate is

$$\lambda(\nu, \xi) = \frac{b(\xi)}{\int_\xi^\infty b(t)dt} .$$

Here again $a_n = o$, $b_n = +\infty$.

EXAMPLE 2.4: Virtual waiting times. (The M/G/1 queue)

Customers arrive at a single-server queue according to Poisson process with rate μ, and have i.i.d. service time requirements with distribution F. The virtual waiting time ξ_t is the time a customer arriving at time t would have to wait for service. Piecewise linear process structure is:

$$\nu_t = \begin{cases} 1 & \text{if queue is not empty} \\ 0 & \text{if queue is empty} \end{cases}$$

and $a_1 = o$, $b_1 = \infty$, $a_o = b_o = o$. Here ξ_t moves to left with uniform speed and transition to (o,o) is certain if $x_{t^-} = (1,o)$.

A more general definition of PL process of Gnedenko and Kovalenko allows ξ_t to move in an open subset O_{ν_t} of $\mathbb{R}^{d(\nu_t)}$ with uniform speed in a fixed direction $V(\nu_t)$. Again transition must take place if $\xi_{t-} \in \partial O_{\nu_t}$, the boundary of O_{ν_t}.

<u>EXAMPLE 2.5</u>: <u>VWT with renewal process arrivals</u>. (The GI/G/1 queue)

Suppose the inter arrival times in Example 2.4 are not exponential, but form a renewal process with inter arrival density $b(.)$. Now the appropriate structure is ν is 0 or 1 as before, $d(1) = 2$, $d(0) = 1$. (When $\nu = 1$ we have to remember both the value of VWT and the time since the last arrival.)

We cannot accommodate this in previous framework, because there $[a_n, b_n]$ is fixed, whereas here the length of the interval is random.

Davis [7] introduced the piecewise deterministic (PD) process which is a further generalization. It is similar to the piecewise linear process, except that ξ_t satisfies some ordinary differential equation, rather than moving in straight line.

<u>EXAMPLE 2.6:</u> <u>Shot noise</u>. This has sample functions similar to the VWT process except that decay between arrivals is exponential rather than linear (fig. 2.1).

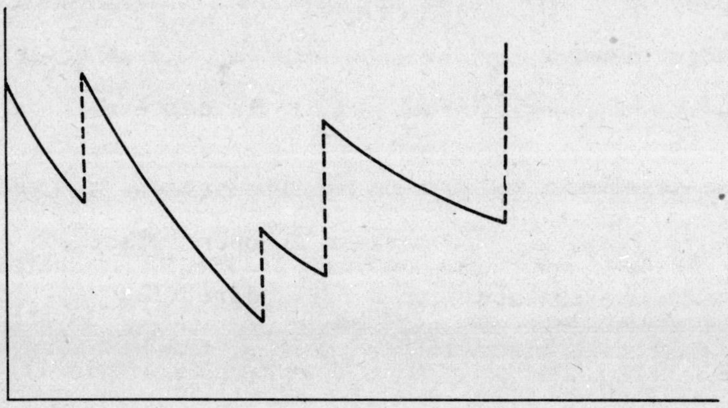

Fig. 2.1

EXAMPLE 2.7: A model for capacity expansion. Suppose that the demand for some utility is monotone increasing and follows a Poisson process with rate μ. Each 'unit' of supply provides q units of capacity. These are built one at a time at a cost of Rs.p. Investment takes place at a rate of Rs.u(t)/week and $u(t) \leq$ constant. When

$$\int_0^t u(s)\, ds = p,$$

then the project is finished, capacity is increased by q and investments are channelled into next project.

Denote d_t = demand; c_t = capacity at time t;
ξ_t = cumulative investment in current project

$$= \int_\tau^t u(s)\, ds,$$

where τ is the last time project was completed. Investment is determined by some "policy" ψ, i.e.,

$$u(t) = \frac{d}{dt}\xi_t = \psi(c_t, d_t, \xi_t)$$

where (c_t, d_t, ξ_t) is the current "situation". Define $\nu_t = (c_t, d_t)$. Then the process $x_t = (\nu_t, \xi_t)$ evolves in the state space $E = Z_+^2 \times [o,p]$ (Z_+^2 is the 2-dimensional positive integer lattice). Then for $\nu = (c,d)$, if $g_\nu(\xi) = \psi(c,d,\xi)$, ξ_t satisfies $\frac{d}{dt}\xi_t = g_{\nu_t}(\xi_t)$.

The piecewise-deterministic process;

Let K be a countable set and $d: K \to \mathbb{N}$ (= natural numbers) be a given function. For each $\nu \in K$, M_ν is an open subset of $\mathbb{R}^{d(\nu)}$ (M_ν can be a $d(\nu)$-dimensional manifold). Then the state space of the PD process is

$$E = \bigcup_{\nu \in K} M_\nu = \{(\nu,\xi): \nu \in K, \xi \in M_\nu\}.$$

Let

$$\mathcal{E} = \{ \bigcup_{\nu \in K} A_\nu : A_\nu \in \mathcal{B}(M_\nu)\}.$$

Then $(E,\mathcal{E})$ is a Borel space. Then the process is $x_t = (\nu_t, \xi_t)$. The probability law of (x_t) is specified by

(i) Vector fields $(X_\nu, \nu \in K)$

(ii) A 'rate' function $\lambda: E \to \mathbb{R}_+$

(iii) A transition measure $Q: \mathcal{E} \times E \to [o,1]$.

Assume that corresponding to each X_ν there is a unique integral curve $\phi_\nu(t,z)$, i.e. $\phi_\nu(t,z)$ satisfies

$$\frac{d}{dt} f(\phi_\nu(t,z)) = X_\nu f(\phi_\nu(t,z))$$

$$\phi_\nu(o,z) = z$$

for every smooth function f, and $\phi(t,z)$ exists for all $t \geq o$. Let ∂M_ν be the boundary of M_ν. $\partial^* M_\nu$ is those points in M_ν at which integral curves exit from M_ν, i.e.

$$\partial^* M_\nu = \{z \in \partial M_\nu : \phi_\nu(t,\xi) = z \text{ for some } (t,\xi) \in \mathbb{R}_+ \times M_\nu\}.$$

Let

$$\Gamma^* = \{(\nu,z): \nu \in K, z \in \partial^* M_\nu\}.$$

So Γ^* is the set of points on the boundary at which jumps may take place. For $x = (\nu,z) \in E$, denote

$$t^*(x) = \inf \{t > o: \phi_\nu(t,z) \in \partial^* M_\nu\}.$$

Write $Xh(x)$ for the function whose value at $x = (\nu, z)$ is $X_\nu h(\nu,.)(z)$. For λ, we suppose that the function $t \to \lambda(\nu,\phi_\nu(t,z))$ is Lebesgue integrable on $[o, \varepsilon[$

for some $\varepsilon > 0$. $Q(.,x)$ is a probability measure on $(E,\mathcal{E})$ for each $x \in E \cup \Gamma^*$.

The motion of the process (x_t) starting from $x = (n,z) \in E$ is described as follows. Define

$$F(t) = \begin{cases} \exp(-\int_0^t \lambda(n,\phi_n(s,z))ds), & t < t^*(x) \\ 0 & , t \geq t^*(x). \end{cases}$$

This is the distribution of T_1, the first jump time. More precisely, $F(t)$ is the survivor function

$$F(t) = P_x[T_1 > t] .$$

Now let Z_1 be an E - valued random variable with distribution $Q(.; \phi_n(T_1,z))$. Then define

$$x_t = \begin{cases} (n, \phi_n(t,z)) & t < T_1 \\ Z_1 & t = T_1 \end{cases}$$

and restart with (n,z) replaced by Z_1. Assume $T_k \uparrow \infty$ a.s. Then x_t defines a measurable mapping from $(\Omega,\mathcal{A},P)$ (countable product of unit interval probability spaces) into the space of right continuous E - valued functions. This defines a measure P_x on the canonical space.

NB: The condition on λ ensures that $T_1 > 0$ a.s. and hence that $T_k - T_{k-1} > 0$ a.s.

PROPOSITION 2.2: (x_t, P_x) is a Markov process.

PROOF: Suppose that $T_k \leq t < T_{k+1}$. The distribution of $T_{k+1} - T_k$ is given by

$$P[T_{k+1} - T_k > s] = \begin{cases} \exp(-\int_o^s \lambda(\nu_{T_k}, \phi_{T_k}(n, \xi_{T_k})du), & s < t^*(x_{T_k}) \\ 0, & s \geq t^*(x_{T_k}). \end{cases}$$

Denote $\nu = \nu_{T_k}$, $\xi = \xi_{T_k}$. Then for $s > t$ and $s < t^*(x_{T_k})$

$$\begin{aligned} P[T_{k+1} > s | T_k, T_{k+1} > t] &= P[T_{k+1} - T_k > s - T_k | T_k, T_{k+1} - T_k > t - T_k] \\ &= \exp[-\int_{t-T_k}^{s-T_k} \lambda(\nu, \phi(u,\xi))du] \\ &= \exp[-\int_o^{s-t} \lambda(\nu_t, \phi_{\nu_t}(u, \xi_t))du] \end{aligned}$$

where we used the semigroup property of ϕ. Since the process "restarts" at T_{k+1}, the law of the process for $s > t$ given part upto t coincides with the law given x_t. Hence the Markov property.

Let $\Gamma \subset \Gamma^*$ be the subset for which, if $y = (\nu, \xi) \in \Gamma^*$

$P[T_1 = t^*(x)] \to 1$ as $x = (n, z) \to y$.

Then Γ is called the "essential" boundary.

<u>EXERCISE 2.2</u>: Prove that $y \in \Gamma$ if and only if $P_x[T_1 = t^*(x)] > o$ for some $x = (\nu, z)$.

So $P_x[x_{s-} \in \Gamma - \Gamma^*$ for some $s > o] = 0$, and $Q(A,x)$ need not be specified for $x \in \Gamma - \Gamma^*$.

<u>EXAMPLE 2.8</u>: Here ν has only a single value; so delete it. ξ takes values in $M = [0,1] \times \mathbb{R}_+$, $\lambda = 0$ and $X = \frac{\partial}{\partial \xi_1}$. Then $\Gamma^* = \{(1,y): y \in \mathbb{R}_+\}$. Let $Q(.;(1,y)) = \delta(1 - \frac{1}{2}y, \frac{1}{2}y)$.

Then starting at $x_o = (0,1)$, we have

$$T_n = \sum_{k=1}^{n} \frac{1}{2^{k-1}} \quad \text{so that} \quad \lim_{n \to \infty} T_n < \infty.$$

The same effect could be achieved by the combined effect of λ and Q, suitably chosen. So, we prefer to assume $T_\infty = \infty$ a.s. rather than stating sufficient conditions on X, λ, Q to ensure this. To illustrate the difference between Γ and Γ^*, suppose that $\lambda(\xi_1, 1) = \frac{1}{1-\xi_1}$ (This is equivalent to saying T_1 is uniformly distributed on $[0,1]$ if the process starts at $(0,1)$). Then $(1,1)$ is never hit, whatever be the starting point. So $(1,1) \in \Gamma - \Gamma^*$.

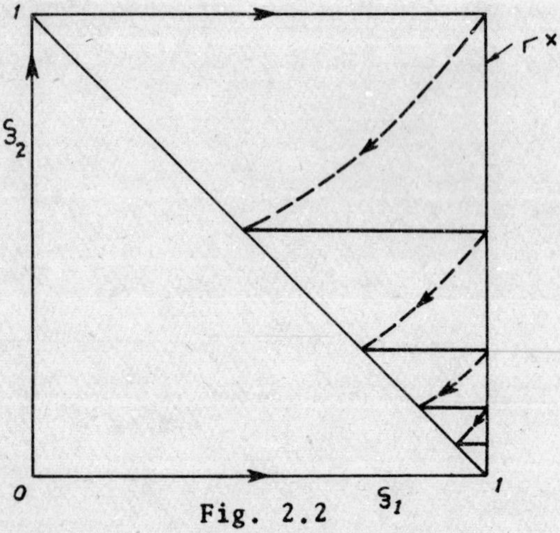

Fig. 2.2

<u>The Associated Jump Process.</u>

Let (x_t) be a PD process. Define the associated jump process (z_t) by

$$z_t = x_{T_k}, \quad t \in [T_k, T_{k+1}[\tag{2}$$

This is an E - valued jump process such that $Z_{T_k} = x_{T_k}$.
Let $F_t = \sigma\{x_s, s \le t\}$ and $F_t^z = \sigma\{z_s, s \le t\}$.

PROPOSITION 2.3: $F_t = F_t^z$ for each t

PROOF: This follows from the fact that there is a one-to-one mapping from $x_{[o,t]}$ to $z_{[o,t]}$. $x \to z$ is given by (2). Conversely, if $z_{[o,t]}$ is given then $x_{[o,t]}$ can be constructed since the motion in the interval $[T_k, T_{k+1}[$ is deterministic.

NB: (1) x_t and z_t are not in one-to-one correspondence at each fixed time t.

(2) (z_t) is not a Markov process.

Since $F_t = F_t^z$, we can apply jump process theory. Define

$$p(t,A) = \sum_{T_i \le t} I_{(x_{T_i} \in A)}$$

$$p_t^* = \sum_{T_i \le t} I_{(x_{T_i^-} \in \Gamma)}$$

$$\tilde{p}(t,A) = \int_o^t Q(A,x_s)\lambda(x_s)ds + \int_o^t Q(A,x_{s-})dp_s^* \tag{3}$$

PROPOSITION 2.4: Suppose $E(p(t,E)) < \infty$. Then for each

$$A \in E,\ q(t,A) = p(t,A) - \tilde{p}(t,A) \tag{4}$$

is an F_t - martingale.

PROOF: From previous results, the compensator of $p(t \wedge T_1, A)$ is

$$\tilde{p}(t \wedge T_1, A) = - \int_{]o,t \wedge T_1]} Q(A, x_{s-}) \frac{dF_s}{F_{s-}}.$$

But

$$F_t = \begin{cases} \exp\left(-\int_0^t \lambda(x_s)ds\right) & t < t_1^*(x) \\ 0 & t \geq t_1^*(x). \end{cases}$$

Thus $-\dfrac{dF_t}{F_t} = \lambda(x_t)\, dt$ for $t < t_1^*(x)$ and

$$\frac{\Delta F_{t_1^*}}{F_{t_1^*-}} = 1.$$

This verifies the result for $t \leq T_1$. As before, we show by considering intervals $]T_{k-1}, T_k]$ that the compensator of $p(t \wedge T_n, A)$ is $\tilde{p}(t \wedge T_n, A)$ given by (3). Since $p(t,A)$ and $\tilde{p}(t,A)$ are monotonic increasing functions and $T_n \uparrow \infty$ a.s. $E(p(t,E)) < \infty$, taking the limit, we have

$$q(t,A) = p(t,A) - \tilde{p}(t,A)$$

is a martingale.

<u>EXERCISE 2.3</u>: Show that p_t^* is an F_t - predictable process.

Then (4) is the Doob-Meyer decomposition of the submartingale p.

The next step is to use stochastic integrals to calculate the extended generator of x_t. Choose the following integrands. For measurable $f: \bar{E} \rightarrow \mathbb{R}$, define

$$Bf(x,s,\omega) = f(x) - f(x_{s-}(\omega))$$

Then $Bf \in L_1(p)$ if

$$E \sum_{T_i \leq t} |f(x_{T_i}) - f(x_{T_i-})| < \infty$$

for each $t \geq o$. This certainly holds if f is bounded and $E\,p(t,E) < \infty$.

$$\int_o^t \int_E Bf(y,s,w)\tilde{p}(ds,dy) = \int_{]o,t]} \int_E (f(y)-f(x_{s-}))Q(dy;x_{s-})\lambda(x_s)ds$$

$$+ \int_{]o,t]} \int_E (f(y)-f(x_{s-}))Q(dy;\, x_{s-})dp_s. \tag{5}$$

Suppose that f satisfies the boundary condition

$$f(x) = \int_E f(y)\, Q(dy;\, x) \quad , \; x \in \Gamma. \tag{6}$$

Then the second integral in (5) is zero. The following result characterizes the extended generator A of (x_t).

THEOREM 2.1: The domain $\mathcal{D}(A)$ of the extended generator A of (x_t) consists of those functions f satisfying

(i) For each $(n,z) \in E$ the function $t \to f(n,\phi_n(n,z))$ is absolutely continuous for $t \in [o,t^*(n,z)[$.

(ii) The boundary condition (6) is satisfied.

(iii) $Bf \in L_1^{loc}(p)$.

Then for $f \in \mathcal{D}(A)$

$$Af(x) = Xf(x) + \lambda(x) \int_E [f(y) - f(x)]Q(dy;x). \tag{7}$$

PROOF: Suppose that f satisfies (i) - (iii). Then $\int Bf\, dq$ is a local martingale, and

$$\int_o^t Bf\, dq = \sum_{T_i \leq t} f(x_{T_i}) - f(x_{T_i^-})$$

$$- \int_o^t \int_E [f(y) - f(x_s)]Q(dy;\, x_s)\lambda(x_s)ds.$$

Now,

$$\sum_{T_i \le t} f(x_{T_i}) - f(x_{T_i^-}) = [\sum_{T_i \le t} (f(x_{T_i}) - f(x_{T_{i-1}}))$$

$$+ f(x_t) - f(x_{T_n})]$$

$$- [\sum_{T_i \le t} (f(x_{T_i^-}) - f(x_{T_{i-1}})) + f(x_t) - f(x_{T_n})]$$

where T_n is the last jump time before t. The first bracket is $(f(x_t) - f(x_o))$. Note that

$$f(x_{T_i^-}) - f(x_{T_{i-1}}) = \int_{T_{i-1}}^{T_i} X_{\nu_{T_{i-1}}} f(\nu_{T_{i-1}}, \phi_{\nu_{T_{i-1}}}(\xi_{T_{i-1}}, s) ds \text{ a.s.}$$

So the second bracket is equal to $\int_o^t X f(x_s) ds$ and

$$\int Bf\, dq = f(x_t) - f(x_o) - \int_o^t \int_E (f(y) - f(x_s)) Q(dy, x_s) \lambda(x_s) ds$$

$$- \int_o^t X f(x_s) ds.$$

So Af is given by (7) and $C_t^f = \int_o^t Bf\, dq$. Conversely, suppose $f \in \mathcal{D}(A)$. Then there exists a function h such that $s \to h(x_s)$ is Lebesgue integrable and $M_t = f(x_t) - f(x_o) - \int_o^t h(x_s) ds$ is a local martingale. By the martingale representation theorem, $M_t = M_t^g$ for some $g \in L_1^{loc}(p)$. Now the jumps of M_t and M_t^g must agree; these only occur when $t = T_i$ for some i and are then given by

$$\Delta M_t = M_t - M_{t-} = f(x_{T_i}) - f(x_{T_i^-}).$$

$$\Delta M_t^g = M_t^g - M_{t-}^g$$

$$= g(x_t, t, w) - \int_E g(y, t, w) Q(dy, x_{t-}) I_{(x_{t-} \in \Gamma)}$$

at $t = T_i$. It follows that

$$g(x,t,w)I_{(x_{t-}\notin\Gamma)} = (f(x)-f(x_{t-}))I_{(x_{t-}\notin\Gamma)}$$

except possibly on a set $G \in E * P$ such that

$$E_y \int_{\mathbb{R}_+ \times E} I_G \, p(dt, dx) = 0 \quad \text{for all} \quad y \in E.$$

Now suppose $x_{T_i^-} = z \in \Gamma$; then

$$f(x) - f(z) = g(x,t,\omega) - \int_E g(y,t,\omega)Q(dy; z)$$

for all x except a set $A \in E$ such that $Q(A,z) = 0$. Since only the first terms on the left and right involve x it must be the case that

$$f(x) = g(x,t,\omega) + \tilde{f}(t,\omega)$$

and

$$f(z) = \int_E g(y,t,\omega) \, Q(dy; z) + \tilde{f}(t,w)$$

for some predictable process $\tilde{f}$. Since $g = f - \tilde{f}$,

$$f(z) = \int_E f(y) \, Q(dy; z)$$

for $z \in \Gamma$, i.e. f satisfies condition (ii). Hence

$$g(x,t,\omega) = f(x) - f(x_{t-}).$$

Hence we get

$$\| (Bf - g) \, I_{(t < \sigma_n)} \|_{L_1(p)} = 0.$$

So condition (iii) is satisfied. Fix ω and consider $(M_t)_{0 \le t < t_1(\omega)}$ starting at (ν_0,ξ_0), then

$$M_t = f(\nu_0,\phi_{\nu_0}(t,\xi_0)) - f(\nu_0,\xi_0) - \int_0^t h(x_s)ds$$

$$M_t^g = \int_0^t \int_E (f(y) - f(x_s))Q(dy;x_s)\lambda(x_s)ds.$$

Hence $f(\nu_0,\phi_{\nu_0}(t,\xi_0))$ is absolutely continuous for $t < T_1(\omega)$.

Since (ν_o, ξ_o) is arbitrary and $T_1(w) > 0$ a.s. this shows that (i) is satisfied.

A "Feynman-Kac" formula.

This is used to calculate expected values of functionals such as

$$E_x[\int_o^t e^{-\alpha s} c(s,x_s)ds + e^{-\alpha t} \Phi(x_t)].$$

There is no extra generality in allowing a P.D. Process to be time-varying, because time can always be included as one component of ξ_t. However, it is sometimes convenient to consider the joint process (t,x_t) with generator $\tilde{A} = \frac{\partial}{\partial t} + A$. Then for $f \in \mathcal{D}(\tilde{A})$

$$f(t,x_t) - f(o,x_o) = \int_o^t (\frac{\partial}{\partial s} + A)f(s,x_s)ds + \int_o^t Bf\, dq.$$

If $(\frac{\partial}{\partial s} + A)\, f(s,x_s) = o$ and $Bf \in L_1(p)$, then $f(t,x_t)$ is a martingale, so it has constant expectation

$$E_{x_o} f(t,x_t) = f(o,x_o).$$

Then

$$f(o,x_o) = E_{x_o} \Phi(x_t)$$

where

$$f(t,x) = \Phi(x) \qquad (\Phi \text{ prescribed}).$$

PROPOSITION 2.5: Let $t > o$ be fixed and $\alpha: [o,t] \times E \to \mathbb{R}_+$, $c: [o,t] \times E \to \mathbb{R}$ and $\Phi: E \to \mathbb{R}$ be measurable functions. Suppose $f: [o,t] \times E \to \mathbb{R}$ satisfies:

$$\left.\begin{array}{ll} \text{(i)} & f(s,.) \in \mathcal{D}(\tilde{A}) \\ \text{(ii)} & f(t,x) = \Phi(x),\ x \in E \\ \text{(iii)} & Bf \in L_1(p) \end{array}\right\} \qquad (8)$$

$$\frac{\partial f(s,x)}{\partial s} + Af(s,x)-\alpha(s,x)f(s,x)+c(s,x) = 0 \qquad (9)$$

$$(s,x) \in [o,t[\times E.$$

Then

$$f(o,x) = E_{o,x}[\int_o^t \exp(-\int_o^s \alpha(u,x_u)du)c(s,x_s)ds + \exp(-\int_o^t \alpha(u,x_u)du)\Phi(x_t)] \qquad (10)$$

PROOF: Suppose f satisfies (8). Define

$$e_s = \exp(-\int_o^s \alpha(u,x_u)du).$$

Then

$$\begin{aligned} d(e_s f(s,x_s)) &= e_s df(s,x_s) + f(s,x_s)de_s \\ &= e_s(\frac{\partial f}{\partial s} + Af)ds + e_s\, Bf\, dq \\ &\quad - \alpha(s,x_s)e_s f\, ds \\ &= -e_s\, c(s,x_s)ds + e_s Bf\, dq \quad \text{(by (9))}. \end{aligned}$$

Now by (iii), $e_s\, Bf \in L_1(p)$ since $e_s \leq 1$. Thus the last term is a martingale and

$$E_x[e_t f(t,x_t)-f(o,x)] = -E_x[\int_o^t e_s c(s,x_s)ds].$$

This with (ii) gives (10).

EXAMPLE 2.9: The Renewal Equation:

Let (N_t) be a renewal process with inter arrival density $f(.)$. Let $m(t) = EN_t$. Since the process "restarts" at renewal times,

$$E[N_t|T_1 = s] = \begin{cases} 0, & s > t \\ m(t-s)+1, & s < t. \end{cases}$$

So

$$m(t) = \int_0^\infty E[N_t | T_1 = s]\, f(s) ds$$

which gives the renewal equation

$$m(t) = \int_0^t (1 + m(t-s))\, f(s)\, ds. \qquad (11)$$

This can be solved by Laplace transforms. Defining

$$\hat{f}(p) = \int_0^\infty e^{-pt} f(t)\, dt$$

etc., we get

$$\hat{m}(p) = \hat{m}(p)\, \hat{f}(p) + \frac{1}{p} \hat{f}(p).$$

So

$$\hat{m}(p) = \frac{\frac{1}{p} \hat{f}(p)}{1-\hat{f}(p)} .$$

In particular, for the Poisson process $f(t) = \lambda e^{-\lambda t}$,

$$\hat{f} = \frac{\lambda}{\lambda + p}$$

will give

$$\hat{m}(p) = \frac{\lambda}{p^2}$$

to get

$$m(t) = \lambda t.$$

<u>EXERCISE 2.4</u>: Compute $M_\tau(t) = E_\tau N_t$, where the component in service at time 0 has age τ (and is replaced by a new component when it fails).

(N_t) is a PD process if we take $x_t = (\nu_t, \xi_t)$ where $\nu_t = N_t$ and ξ_t is the time since last renewal. Then

$$X_\nu = \frac{\partial}{\partial \xi}, \qquad \lambda(\xi) = \frac{f(\xi)}{\int_\xi^\infty f(u) du}$$

and $Q(.;\nu,\xi) = \delta_{(\nu+1,0)}$, so that

$$Af(\nu,\xi) = \frac{\partial}{\partial\xi} f(\nu,\xi) + \lambda(\xi)[f(\nu+1,0) - f(\nu, 0)].$$

Use proposition 2.5 with $\alpha = c = 0$ and $\Phi(x) = \nu$ to get

$$f(0,\nu,\xi) = E_{(\nu,\xi)} \nu_t.$$

Clearly

$$f(s, \nu+1,\xi) = f(s,\nu,\xi) + 1.$$

Define

$$f(s,0,\xi) = h(s,\xi).$$

Then the equation for f(or h) becomes

$$\frac{\partial}{\partial s} h(s,\xi) + \frac{\partial}{\partial\xi} h(s,\xi) + \lambda(\xi)[1+h(s,0)-h(s,\xi)] = 0 \quad (12)$$

$$h(t,\xi) = 0.$$

Define

$$z(u) = h(u,u).$$

Then

$$\frac{d}{du} z(u) = - \lambda(u)[1 + h(u,0) - z(u)]$$

$$z(t) = 0.$$

Thus $z(u)$ satisfies

$$\dot{z}(u) = \lambda(u)\, z(u) - \lambda(u)\, [1 + h(u,0)] \quad (13)$$

where

$$\lambda(u) = - \frac{\dot{F}}{F} = \frac{f}{F}, \quad F(u) = \int_u^\infty f(s)\, ds.$$

Equation (13) is a linear ODE satisfied by $z(u)$.

The transition function corresponding to $\lambda(u)$ is

$$\Phi(u,v) = \frac{F(u)}{F(v)}$$

Hence (13) has the following solution at time 0:

$$z(o) = h(o,o) = \int_o^t f(u)[1 + h(u,o)]du \qquad (14)$$

Define

$$m(s) = h(t - s,o).$$

Then (14) coincides with the renewal equation (11). Having determined $h(u,o)$ $o \leq u \leq t$, $h(s,\xi)$ for $s \neq \xi \neq o$ can be calculated from (12). The result will be equivalent to that of Exercise 2.4.

CHAPTER II

OPTIMAL CONTROL OF PD PROCESSES

General formulationsof stochastic control problems have been studied using martingale theory, where the conditions for optimality, existence of optimality are derived (El Karoui [15]). But this does not give ways of computing optimal control. Control of Markov jump processes has been studied using dynamic programming (Pliska [21]). In this Chapter, we will be dealing with control theory for PD processes, following Vermes [25].

Let Y be a compact metric space. Control arises when the system functions X, λ, Q contain a parameter $y \varepsilon Y$ i.e., for $x = (\nu,\xi)$

$$X^y f(x) = \sum_i b(\nu,\xi,y) \frac{\partial f(\nu,\xi)}{\partial \xi_i}$$

$$Q = Q(A, x,y).$$

A feedback <u>policy</u> (or <u>strategy</u>) is a function $u: \mathbb{R}_+ \times E \to Y$. Let $\mathcal{U}$ denote the set of all strategies. u is <u>stationary</u> if there is no t - dependence, i.e., $u: E \to Y$. Corresponding to policy u we get a PD process with characteristics X^u, λ^u, Q^u given by

$$X^u f = \sum_i b(\nu,\xi, u(x)) \frac{\partial f}{\partial \xi_i}(x)$$

$$\lambda^u(x) = \lambda(x, u(x))$$

$$Q^u(A;x) = Q(A;x,u(x)).$$

More conditions on u will be added when required. Then we get a PD process x_t with probability measure P^u determined by X^u, λ^u, Q^u.

Given a cost function, say, for example,

$$J_x(u) = E_x^u\left[\int_o^t e^{-\alpha s}\, c(x_s,u_s)ds + e^{-\alpha t}\Phi(x_t)\right]$$

where E_x^u is the expectation w.r.t. P^u starting at x and $\alpha > o$. The control problem is to choose $u(.)$ to minimise $J_x(u)$. The "usual" approach to such problems is via "dynamic programming". Let $V(s,x)$ be a function of (s,x). Introduce the Bellman-Hamilton-Jacobi equation

$$\frac{\partial V(s.x)}{\partial s} + \min_{y \in Y}\,[A^y V(s,x) + c(x,y)] - \alpha V(s,x) = 0 \quad (B)$$

where A^y is the generator corresponding to X^y, $\lambda(.,y)$, $Q(.,y)$.

If Y has one point, then this coincides with the equation for J_x as before.

PROPOSITION 1: Suppose (B) has a "nice" solution (i.e., satisfies boundary condition etc.). Then

$$V(o,x) = \min_{u \in \mathcal{U}} J_x(u)$$

and the optimal strategy $u^o(s,x)$ satisfies

$$A^{u^o(s,x)}\, V(s,x) + c(x,u^o(s,x)) = \min_{y \in Y}\,(A^y\, v+c).$$

PROOF: Same calculations arise as before. Let x_t correspond to an arbitrary control policy u. Then

$$d(e^{-\alpha s}V(s,x_s)) = -\alpha e^{-\alpha s}V(s,x_s)ds + e^{-\alpha s}\left(\frac{\partial V}{\partial s} + A^u V\right)ds +$$

$$+ e^{-\alpha s}BV\, dq \geq -e^{-\alpha s}c(x_s,u(s,x_s))ds + e^{-\alpha s}Bu\, dq \quad (1)$$

So

$$V(o,x) \leq E_x[\int_o^t e^{-\alpha s} c(x_s,u_s)ds + e^{-\alpha t} \Phi(x_t)]$$
$$= J_x(u).$$

Now suppose $u = u^o$, then "equality" holds in place of "inequality" in (1). So

$$V(o,x) = J_x(u_o).$$

So u^o is optimal.

Objections: (1) There is no general theory under which (B) has a solution.

(2) $u^o(x)$ constructed as above may fail to be an admissible control: to make sense of it, we must be able to solve the ODE

$$\frac{d}{ds}\xi(s) = b_\nu^{u^o}(\xi_s) = b(\nu,\xi,u^o(s,\xi)).$$

There is no guarantee that u^o leads to a "solvable" ODE.

So we must redefine "admissible controls" so that this is avoided.

REMARK 1: In control of diffusion processes, the equation is

$$dx_t = b(x_t, u(x_t))dt + \sigma(x_t)dW_t.$$

Here we "handle" nonsmooth u by using weak solutions.

REMARK 2: In deterministic control, one uses open-loop controls depending only on time. The equation here is of the form

$$\dot{x} = b(x_t, u(t)).$$

Then solution is well defined for the measurable $u(.)$.

Special cases: (1) Control only appears in Q. Then the problem reduces to a sequential decision problem where a "decision" is taken each time a jump occurs. (Rosberg, Varaiya and Walrand [22]).

(2) $X = o$. Here Markov jump process with piecewise constant paths are considered. Control appears in λ and Q. Then

$$A^u f = \lambda(x,u(x))(\int_E (f(z)-f(x))Q(dz;x,u(x))$$

is a bounded operator on $\mathcal{B}(E)$. Regard (B) as an ODE in Banach space $\mathcal{B}(E)$. Let $V(s) := V(s,.)$, then

$$\frac{dV}{ds} = g(V(s)) = \min_{y \in Y} (A^y V + c).$$

So g is a nonlinear function, but it is Lipschitz continuous in V [Pliska [21]].

(3) Piecewise linear processes (Vermes [25]). Here ξ_t is one dimensional and $X = \frac{\partial}{\partial \xi}$. Control appears in λ and Q.

Consider a 'stationary' control problem, where the Bellman equation takes the form

$$\min_{y \in Y} (A^y V + c(x,y)) = 0$$
$$V(x) = \Phi(x) , \quad x \in E_T.$$

This corresponds to minimising

$$E_x(\int_0^\tau c(x_s,u_s)ds + \Phi(x_\tau)),$$

where τ is the first hitting time of some target set E_T. Then

$$A^y V(x) = \frac{\partial}{\partial \xi} V(\nu,\xi) + \lambda(x,y) \int_E (V(z)-V(x))Q(dz,x,y).$$

Suppose $\nu \varepsilon \{1,2,\ldots,N\}$ and

$$V(\xi) = \begin{bmatrix} V(1,\xi) \\ \\ V(N,\xi) \end{bmatrix}$$

Then Bellman equation takes the form

$$\frac{d}{d\xi} V(\xi) = g(V(.)).$$

This is an "ordinary" functional differential equation with non-standard boundary condition. Vermes showed existence of an optimal feedback strategy in special cases.

'Generalized' Dynamic Programming Conditions:

Let us consider next optimal control of the deterministic differential system:

$$\dot{x}_t = f(x_t,t,u_t) \ , \ t \ \varepsilon \ [t_o,t_1]. \qquad (2)$$

Then the control problem is to

$$\text{minimize} \int_{t_o}^{t_1} \ell(x_t,t,u_t)dt$$

over "admissible" control/trajectory pairs u_t, x_t i.e., pairs of functions for which

(i) (2) is satisfied,

(ii) $x(t_1) = x_1$, $x(t_o) = x_o$ with x_o, x_1 given,

(iii) $x_t \ \varepsilon \ \bar{A}$, $u_t \ \varepsilon \ \Omega$, where $\bar{A}$, Ω are compact and $A = \bar{A} \times [t_o, t_1]$.

This will be called the strong problem (S).

We assume (a) an admissible pair (x_t,u_t) exists, and also we make a temporary assumption

(b) $\begin{bmatrix} f(x,t,\Omega) \\ \\ \ell(x,t,\Omega) \end{bmatrix}$ is convex.

This enables "relaxed controls" to be avoided. Define

$\eta(S)$ = "value" of S

i.e., $\inf_{(x_t,u_t)ad} \int \ell\, dt.$

THEOREM 1: There exists an optimal admissible pair (x_t, u_t) for the strong problem.

This is a "standard" result in optimal control theory (Vinter and Lewis [26]). It depends critically on the convexity assumption (b).

A Sufficient Condition for Optimality: (Standard Dynamic Programming). Suppose (x_t, u_t) is admissible and ϕ is in $C^1(A)$ such that

$$\phi_t(t,x) + \max_{u \in \Omega} (\phi_x(x,t)f(x,t,u)-\ell(x,t,u)) = 0$$
$$(x,t,u) \in A \times \Omega$$

$$\phi(x_1, t_1) = 0$$

and

$$\phi_t(t,x_t) + \phi(x_t,t)f(x_t,t,u_t) - \ell(x_t,t,u_t) = 0 \quad \text{a.a.t.}$$

Then (x_t,u_t) is optimal and $\eta(S) = -\phi(x_o,t_o)$. The main result of Vinter and Lewis is as follows.

THEOREM 2: The strong problem has a solution (i.e. there exists an optimal pair (x_t,u_t)). There exists a sequence $\{\phi^i\}$ in $C^1(A)$ such that

$$\phi_t^i + \max_{u \in \Omega} (\phi_x f - \ell) \leq o, \quad (x,t) \in A$$

$$\phi^i(x_1, t_1) = 0$$

<u>and</u> (x_t, u_t) <u>is optimal if and only if</u>

$$\lim_{t \to \infty} H^i(t) = 0 \quad \text{in} \quad L_1[t_o, t_1]$$

<u>where</u>

$$H^i(t) = \phi_t^i(x_t,t) + \phi_x^i(x_t,t)f(x_t,t,u_t) - \ell(x_t,t,u_t).$$

<u>The Weak Problem</u>

For (x_t, u_t) admissible, define $\mu_{x,u} \in C^*(A \times \Omega)$, the dual of $C(A \times \Omega)$, by

$$< g, \mu_{x,u} > = \int_{t_o}^{t_1} g(x_t, t, u_t)\, dt$$

for arbitrary $g \in C(A \times \Omega)$. $\mu_{x,u}$ satisfies

(i) $\mu_{x,u} \in P^+$ (i.e., if $g \geq o$ then $< g, \mu_{x,u} > \geq o$).

(ii) Take $\phi \in C^1(A)$ and

$$g(x,t,u) = \phi(x,t) + \phi_x(x,t)\, f(x,t,u)$$

then

$$< \phi_t + \phi_x f, \mu_{x,u} > = \phi(x_1, t_1) - \phi(x_o, t_o).$$

Define

$$\mathcal{M} = \{\mu \in C^*(A \times \Omega): \text{ (i) and (ii) are satisfied}\}.$$

<u>PROPOSITION 2</u>: $\mathcal{M}$ <u>is weak* compact and convex.</u>

<u>NOTE 1</u>: The cost function for (x_t, u_t) is $< \ell, \mu_{x,u} >$.

Weak Problem (W): Minimise $< \ell, \mu >$ over $\mu \in M$.

So $\eta(W) \leq \eta(S)$.

THEOREM 3: $\eta(S) = \eta(W)$. There exists an optimal x,u for S, so $\mu_{x,u}$ is optimal for W.

Now we incorporate the constraints on μ into the cost function in the following way. Define extended real-valued functions p, q on $C^*(A \times \Omega)$ as follows:

$$p(\mu) = \begin{cases} < \ell, \mu > & \text{if } |\mu| \leq t_1 - t_o,\ \mu \in P^+ \\ +\infty & \text{otherwise.} \end{cases}$$

Let $M_2 = \{\mu \in C^*\text{: Condition (ii) is satisfied}\}$. Then

$$q(\mu) = \begin{cases} 0 & \text{if } \mu \in M_2 \\ -\infty & \text{otherwise .} \end{cases}$$

PROPOSITION 3: p is ℓ.s.c. and convex,

q is u.s.c. and concave,

and

$$\eta(W) = \inf_{\mu \in C^*} \{p(\mu) - q(\mu)\}.$$

The Fenchel dual problem is as follows:

$$\text{(D)} \qquad \max_{\xi \in C(A \times \Omega)} (q^*(\xi) - p^*(\xi))$$

where p^*, q^* are "dual" functionals defined by

$$p^*(\xi) = \sup_{\mu \in C^*} < \xi, \mu > - p(\mu).$$

$$q^*(\xi) = \inf_{\mu \in C^*} < \xi, \mu > - q(\mu).$$

PROPOSITION 4:

$$p^*(\xi) = \max_{(x,t,u)\, \varepsilon\, A \times \Omega} (\xi(t,x,u) - \ell(t,x,u))^+ \times (t_1 - t_o).$$

where $a^+ := \max(a,0)$.

SKETCH OF PROOF:

$$p^*(\xi) = \sup_{\substack{|\mu| \leq t_1 - t_o \\ \mu\, \varepsilon\, P^+}} [< \xi, \mu > - < \ell, \mu >]$$

$$= \sup_{\substack{|\mu| \leq t_1 - t_o \\ \mu \varepsilon\, P^+}} [< \xi - \ell, \mu >] .$$

If $< \xi, \mu > - < \ell, \mu >$ is negative, then the optimum is zero. If $< \xi, \mu > - < \ell, \mu > \geq o$, then put Dirac measure $\times (t_1 - t_o)$ on maximum point to get the result.

PROPOSITION 5 : Let

$$W = \{\xi \;\varepsilon\; C: \xi = \phi_t + \phi_x f \;\; \underline{\text{for some}} \;\; \phi \;\varepsilon\; C^1 (A)\} .$$

Then

$$q^*(\xi) = \begin{cases} -\infty & \text{if } \xi \notin \bar{W} \\ \lim_i (\phi^i(x_1,t_1) - \phi^i(x_o, t_o)) & \text{if } \xi \;\varepsilon\; \bar{W} \end{cases}$$

where

$$\xi = \lim_i \xi^i \quad \underline{\text{and}} \quad \xi^i = \phi^i_t + \phi^i_x f .$$

PROOF: For $\xi \;\varepsilon\; W$, by definition of q and q^*, we get

$$q^*(\xi) = \phi(x_1, t_1) - \phi(x_0, t_o).$$

A similar argument gives the result for $u \;\varepsilon\; \bar{W}$. For $\xi \notin \bar{W}$, there exists a separating hyperplane, i.e., $\bar{\mu} \;\varepsilon\; C^*$ such that

$< \bar{\xi}, \bar{\mu} > \neq o$, $\bar{\xi} \in \bar{W}$ and $< \xi, \bar{\mu} > = o$. If $\mu \in M_2$, then $\mu + c\,\bar{\mu} \in M_2$. So

$$q^*(\xi) = \inf_{\mu \in M_2} < \xi, \mu > = -\infty.$$

Characterizing the solution of (D):

$$\eta(D) = \max_{\xi \in \bar{W}} [\lim_{i} (\phi^i(x_1,t_1) - \phi^i(x_o,t_o))$$

$$- \max_{(x,t) \in A \times \Omega} (\xi(t,x,u) - \ell(t,x,u)^+(t_1 - t_o)]$$

$$= \sup_{\phi \in C^1} (\phi(x_1,t_1) - \phi(x_o,t_o) - \max_{x,t,u} (\phi_t + \phi_x f - \ell)^+(t_1 - t_o)).$$

It is no restriction to assume $\phi(x_1,t_1) = 0$. Then Vinter and Lewis show by an ingenious argument that

$$\eta(D) = \sup (-\phi(x_o,t_o))$$

where the supremum over $\phi \in C^1$ such that $\phi(x_1,t_1) = 0$ and

$$(\phi_t + \phi_x f - \ell) \leq 0 \quad \forall \quad (x,t,u).$$

<u>THEOREM 4</u>: $\eta(D) = \eta(W) = \eta(S)$.

<u>PROOF</u>: This follows from a "standard" result in duality theory: q^* is finite at some point in its domain where p^* is continuous.

Proofs of the main results now follow easily. The strong problem has a solution since $\eta(S) = \eta(D)$.

$$\eta(S) = \lim (- \phi^i(x_o,t_o))$$

for some sequence of ϕ^i's satisfying the "Bellman inequality". The characterization of optimal pairs (x_t, u_t) follows.

<u>REMARK 3</u>: If the set

$$\left\{ \begin{array}{l} f(x, t, \Omega) \\ \ell(x, t, \Omega) \end{array} \right\}$$

is not convex, then the results are still valid but <u>relaxed controls</u> must be used.

A relaxed control μ_t is a $C^*(\Omega)$ - valued function on $[t_o, t_1]$ such that μ_t is a probability measure for every t and $t \to \int g(t,u)\, \mu_t(du)$ is measurable for every continuous function g.

<u>Interpretation</u>: x_t, μ_t is an admissible pair, whenever

$$\frac{dx_t}{dt} = \int_\Omega f(x_t, t, u)\, \mu_t(du);$$

the cost is

$$\int_{t_o}^{t_1} \int_\Omega f(x_t,t,u)\, \mu_t(du)\, dt.$$

<u>Optimal Control of PD Processes</u> (Vermes [25])

In this section, we adopt a slightly modified definition of the PD process (x_t). It will take values in E, a closed subset of $\mathbb{R}^d$, and we suppose that

$$E = E_o \cup E_\partial \cup E_T \qquad \text{(disjoint)}$$

where E_T is a closed set, E_o is an open set and

$$E_\partial = (\bar{E}_o - E_o) - E_T.$$

Let E_o', E_∂', E_T' be compactification of E_o, E_∂, E_T respectively and E' be the disjoint union of E_o', E_∂' and E_T'. Then a controlled PD process is determined by functions

$$f: E_o' \times Y \to \mathbb{R}^d;$$

$$\lambda: E_o' \times Y \to \mathbb{R}_+,$$

and

$$Q: (E_o' \cup E') \times Y \to m_1(E_o)$$

where $m_1(E_o)$ is the set of probability measures on E_o and Y is a compact (control) space. Function f gives the deterministic motion by

$$\dot{x}_t = f(x_t, y_t).$$

We assume f satisfies a Lipschitz condition in x.

Admissible Controls: Feed back controls $u(t) = u(x_t)$ are not the "right" class of controls because the equation $x = f(x, u(x))$ only has a unique solution under strict conditions on $u(.)$. Let

$$\alpha_t = \text{last jump time before } t.$$

$$n(t) = x_{\alpha(t)}$$

$$z(t) = t - \alpha(t).$$

Then $n(t)$, $z(t)$ determine x_t; in fact, for fixed $y \in Y$,

$$x_t = X_{n(t), z(t)}$$

where

$$X_{n,z} = n + \int_0^z f(X_{n,s}, y)ds.$$

Then admissible controls are Y - valued measurable functions $u(n(t), z(t))$. By Caratheodory's theorem, the equation

$$X_{n,z} = n + \int_0^z f(X_{n,s}, u(n,s))\, ds$$

has a unique solution, and PD process is well defined for such u. We will consider the three component process (x_t, z_t, n_t) for notational convenience.

Relaxed controls are functions $\mu: E \times \mathbb{R}_+ \to m_1(Y)$ such that $(n,z) \to \int \phi(n,z,y)\, \mu(dy; n,z)$ is a measurable function for (n,z) for all continuous ϕ. Corresponding to μ, define

$$f^{\mu}(x, n, z) = \int f(x, y)\, \mu(dy, n, z)$$

$$\lambda^{\mu}(x, n, z) = \int \lambda(x, y)\, \mu(dy; n, z)$$

$$Q^{\mu}(A,x,n,z) = \int Q(A, x, y)\, \mu(dy, n, z).$$

Then we construct a PD process (x_t, n_t, z_t) corresponding to f^{μ}, λ^{μ}, Q^{μ} in the usual way.

The strong problem is to minimise $J_{\hat{x}_o}(\mu)$ over admissible relaxed controls μ, where $\hat{x}_o = (x,x,o)$, and

$$J_{\hat{x}_o}(\mu) = E^{\mu}_{x_o}\Big(\int_o^{\tau}\int_Y \ell_o(x_t,y)\mu(dy; n_t,z_t)dt$$

$$+ \sum_{\{t:x_{t-} \in E_{\partial}\}} \int \ell_o(x_{t-},y)\mu(dy;n_{t-}, z_{t-}) + \ell_T(x_{\tau})\Big).$$

Here τ is the first hitting time of set E_T.

MAIN RESULTS:

THEOREM 5: There exists an optimal (relaxed) control.

THEOREM 6: The value function $\psi(x) = \sup \Phi(x,x,o)$ where the supremum is over all functions $\Phi \in C^1(E)$ such that

$$\Phi_z(x,n,z) + \min_{y \in Y} (\nabla_x \Phi(x,n,z) f(x,y)$$

$$+ \lambda(x,y)(\int \Phi(\xi,\xi,o) Q(d\xi,x,y) - \Phi(x,n,z))$$

$$+ \ell_o(x,y)) \qquad (x,n,z) \in \tilde{E}_o \qquad (3)$$

$$\geq 0$$

$$\Phi(x,n,z) \leq \min_{y \in Y} \{\int \Phi(\xi,\xi,o) Q(d\xi,y,x) + \ell_o(x,y)\}$$

$$(x,n,z) \in \tilde{E}_\partial \qquad (4)$$

$$\Phi(x,n,z) \leq \ell_T(x) \ , \ x \in E_T \qquad (5)$$

and $\tilde{E}$ is the space of triplets (x,n,z).

THEOREM 7: There exists a sequence Φ^k satisfying (3), (4), (5) above such that μ^o is optimal if and only if

$$\Phi^k(x,n,z) + \int_Y \{\nabla_x \Phi^k(x,n,z) f(x,y) + \lambda(x,y)[\int \Phi^k(\xi,\xi,o)$$

$$\times Q(d\xi,x,y) - \Phi^k(x,n,z)] + \ell_o(x,y)\ \mu^o(dy,n,z)$$

$$\to 0 \quad \text{in} \quad L_1(Q_o^o). \qquad (6)$$

$$\int_Y \{\int_{E_o} \Phi^k(\xi,\xi,o) Q(d\xi,x,y) + \ell_o(x,y)\} \mu^o(dy,n,z)$$

$$- \Phi^k(x,n,z) \to 0 \quad \text{in} \quad L_1(Q_\partial^o) \qquad (7)$$

$$\Phi^k(x,n,z) - \ell_T(x) \to 0 \quad \text{in} \quad L_1(Q_T^o). \qquad (8)$$

The measures Q_o^o, Q_∂^o and Q_T^o are defined as follows. Denote $\tilde{x}_t = (x_t, n_t, z_t)$. For $A \in \tilde{E}_o$,

$$Q_o^o(A) = \tilde{E}_{\hat{x}_o}^{\mu_o} \int_o^\tau \psi_A(\tilde{x}_t)\, dt$$

<u>which is a measure on</u> $\tilde{E}_o$ <u>and is called potential measure of</u> $\tilde{x}_t$.

$$Q_\partial^o(A) = \tilde{E}_{\hat{x}_o}^{\mu_o} \sum_{t \leq \tau} \psi_A(\tilde{x}_{t-})$$

<u>where</u> $A \in \tilde{E}_\partial$.

$$Q_T^o(A) = \tilde{P}_{\hat{x}_o}^{\mu_o} [\tilde{x}_T \in A]$$

<u>for</u> $A \in \tilde{E}_T$.

Comparing with deterministic case, the necessary and sufficient condition there was that (x_t, μ_t) is optimal if and only if

$$\phi_t^i(x_t,t) + \int \{\phi_x^i(x_t,t)f(x_t,t,u) - \ell(x_t,t,u)\}\mu_t(du)$$

$$\to 0 \quad \text{in } L_1(t_o, t_1).$$

The "probability measure" corresponding to μ_t is Dirac measure on $x(.)$ and $Q_o^o(A)$ is the time spent by $x(.)$ in A. Thus the conditions stated are a direct generalization of the deterministic ones.

<u>REMARK 4</u>: Note that if we define

$$Q_o^o(A) = E_{\tilde{x}} \int_o^\tau \psi_A(\tilde{x}_s)ds$$

then for any positive measurable function g,

$$E_{\tilde{x}} \int_o^\tau g(\tilde{x}_s)ds = \int_{\tilde{E}_o} g(\xi)\, Q_o^o(d\xi);$$

for, if

$$g(\tilde{x}) = \sum_i c_i \, \psi_{A_i}(\tilde{x})$$

then

$$E_{\tilde{x}} \int_0^{\tau} g(\tilde{x}_s)ds = \sum_i c_i \, E_x \int_0^{\tau} \psi_{A_i}(\tilde{x}_s)\, ds$$

$$= \sum_i c_i \, Q_0^{\circ}(A_i)$$

$$= \int_{E_0} g(\xi)\, Q(d\xi).$$

The general case follows by monotone convergence.

REMARK 5: The Q_i° are "potentials of additive functionals" ℓ_t is an additive functional if $\ell \geq 0$ and*

$$\ell_{t+s} = \ell_t + \ell_s \circ \theta_t$$

t, p_t^*, $I_{(t \geq \tau)}$ are some examples of additive functionals.

The potential of an additive functional is an operator

$$U_\ell \, g(\tilde{x}) = E_{\tilde{x}} \int_0^{\tau} g(\tilde{x}_s)d\ell_s.$$

Here Q_0°, Q_∂°, Q_T° correspond precisely to this with $\ell_t = t$, p_t^*, $I_{(t > T)}$ respectively.

The Weak Problem:

The deterministic weak problem involved the fact that

$$\phi(x_1,t_1) - \phi(x_0,t_0) = \int_{t_0}^{t_1} (\phi_t + \phi_x f)ds$$

for any $\phi \in C^1$. The stochastic equivalent of this is Dynkin's formula. To get this in the appropriate form, define operators A^y, B^y as follows.

* θ_t is the shift operator on the space of right continuous functions: $(\theta_t w)_s = \omega_{t+s}$

$$A^y\phi(x,n,z) = \phi_z(x,n,z) + \nabla_x\phi(x,n,z)f(x,y)$$

$$+ \lambda(x,y) \int_{\tilde{E}_o} (\phi(\xi,\xi,o) - \phi(x,n,z))Q(d\xi,x,y)$$

and

$$B^y\phi(x,n,z) = \int_{E_\partial} \phi(\xi,\xi,o)Q(d\xi,x,y) - \phi(x,n,z)$$

for $(x,n,z) \in \tilde{E}_\partial$. Then the Dynkin formula on the interval (o,τ) is

$$\tilde{E}^\mu_{\hat{x}}\, \phi(x_\tau,n_\tau,z_\tau) - \phi(\hat{x})$$

$$= E^\mu_{\hat{x}}[\int_o^\tau \int_Y A^y\phi(x_t,n_t,z_t)\mu(dy;\ n_t,z_t)dt$$

$$+ \int_o^\tau \int_Y B^y\phi(x_t,n_t,z_t)\mu(dy,z_t,n_t)dp^*_t]$$

$$= \int_{\tilde{E}_o} \int_Y A^y\phi(x,n,z)\mu(dy;n,z)Q^\mu_o(dx,dn,dz)$$

$$+ \int_{\tilde{E}_\partial} \int_Y B^y\phi(x,n,z)\mu(dy;x,z)Q^\mu_o(dx,dn,dz)$$

Now

$$\tilde{E}^\mu_{\hat{x}}\phi(x_\tau,n_\tau,z_\tau) = \int_{\tilde{E}_T} \phi(x,n,z)Q^\mu_T(dx,dn,dz).$$

So we can express the Dynkin formula as follows:

$$\phi(\hat{x}_o) = \int_{\tilde{E}\times Y} L\phi(\tilde{x},y)M^\mu(d\tilde{x},dy)$$

where

$$L\phi(x,n,z,y) = \psi_{\tilde{E}_T}\phi(x,n,z)$$

$$+ \psi_{\hat{E}_o} A^y\phi(x,n,z) + \psi_{\tilde{E}_\partial} B^y(x,n,z).$$

$$M^{\mu}(S_1 \times S_2) = Q_T^{\mu}(S_1 \cap \tilde{E}_T)$$

$$+ \int_{S_1 \cap \tilde{E}_o} \int_{S_2} \mu(dy;n,z) Q_o^{\mu}(dx,dn,dz)$$

$$+ \int_{S_1 \cap \tilde{E}_\partial} \int_{S_2} \mu(dy,n,z) Q_\partial^{\mu}(dx,dn,dz).$$

The cost for the relaxed control μ is

$$J_{\hat{x}_o}(\mu) = \int_{E \times Y} \ell(\tilde{x},y) M^{\mu}(d\tilde{x},dy).$$

The following supplementary assumption is required.

$$\inf_{u \in \mathcal{U}} J_{\hat{x}_o}(\mu) = \inf_{\mu \in \mathcal{U}_c} J_{\hat{x}_o}(\mu)$$

for some $c > o$ and $\mathcal{U}$ is the set of relaxed controls,

$$\mathcal{U}_c = \{\mu \in \mathcal{U} : \mu \in [\tau + p_\tau^*] \leq c\}$$

with this assumption the weak problem is to minimize $\int_{\tilde{E} \times Y} \ell dM$ over measures $M \in m_{1+c}(\tilde{E} \times Y)$ (where m_a is the set of positive measures of total mass less than or equal to a) such that

1. $M = M_o + M_\partial + M_T$

where

$M_T \in m_1(\tilde{E}_T)$.

$M_\partial \in m(\tilde{E}_\partial \times Y)$

$M_o \in m(\tilde{E}_o \times Y)$.

2. $\Phi(\hat{x}_o) = \int L\,\phi\, dM, \quad \Phi \in C^1(\tilde{E})$.

From this point on, the development follows the Vinter-Lewis argument closely. We reformulate the weak problem as a convex optimization problem by incorporating the constraints in the cost function and obtain the characterization of optimality by studying the dual problem. The reader is referred to Vermes [25] for the details.

REMARK 6: The optimality condition involves the measures Q_o^o, Q_∂^o, Q_T^o corresponding to μ^o. These can be computed from the following system of equations.

$$A^{\mu^o} h(\tilde{x}) + \psi_{\Gamma \cap E_o} = 0 \ , \ \tilde{x} \in \tilde{E}_o$$

$$B^{\mu^o} h(\tilde{x}) + \psi_{\Gamma \cap \tilde{E}_\partial} = 0 \ , \ \tilde{x} \in \tilde{E}_\partial$$

$$h(\tilde{x}) = \psi_{\Gamma \cap \tilde{E}_T}(\tilde{x}) \ , \ \tilde{x} \in \tilde{E}_T \ .$$

Then

$$Q^o(\Gamma) = h(\hat{x}_o).$$

EXAMPLE 1: If $\Gamma \subset \tilde{E}_o$, then Dynkin's formula says

$$h(\hat{x}_o) = E_{\hat{x}_o}^{\mu^o} \int_o^\tau \psi_\Gamma(\tilde{x}_s)ds$$

$$= Q_o^o(\Gamma).$$

The results outlined above are the first general results on optimal control of PD processes. Obviously much work remains to be done; natural next steps would be to determine necessary conditions for optimality of Pontrjagin type; to develop computational methods; and to study optimal stopping and "impulse control" for PD processes. For some related work, see van der Duyn Schouton [29], Yushkevich [30] and Rishel [31].

PART II FILTERING THEORY

§0. INTRODUCTION

Suppose $\{x_t\}$ is a signal process which represents the state of a system, but cannot be observed directly. We observe a related process $\{y_t\}$. Our aim is to get an expression for the "best estimate" of x_t, given the history of $\{y_t\}$ upto time t.

In Section 1, we give quick derivations of the "Kalman filter" for the linear systems, and nonlinear filtering equations, that of Fujisaki, Kallianpur and Kunita and Zakai's equation for unnormalized conditional density (Kallianpur [19], Davis and Marcus [8]). In Section 2, we will study pathwise solutions of differential equations. In Section 3, we will study the "Robust" theory of filtering as developed by Clark [5], Davis [10] and Pardoux [20]. Here the above filtering equations are reduced to quasi-deterministic form and solved separately for each observation sample path. Also,we will look here into some more general cases of filtering developed by Kunita [17], where the existence of conditional density functions is proved using methods related to the theory of "stochastic flows".

§1. LINEAR AND NONLINEAR FILTERING EQUATIONS

Kalman Filter (Davis [11])

Suppose that the "signal process" x_t satisfies the linear stochastic differential equation

$$dx_t = Ax_t\,dt + c\,dV_t \tag{1}$$

where V_t is some Wiener process. "Observation" y_t is given by

$$dy_t = Hx_t\,dt + dW_t \tag{2}$$

where W_t is a Wiener process independent of V_t. Assume $x_o \sim N(o,P_o)$. To get a physical model, suppose we write (2) as

$$\frac{dy_t}{dt} = Hx_t + \frac{dW_t}{dt}$$

then $\frac{dW_t}{dt}$ corresponds to white noise and $\frac{dy_t}{dt}$ is the "physical" observation.

The filtering problem is to calculate 'best estimate' of x_t given $(y_s,\ s \leq t)$. There are two formulations for Kalman filter.

(a) Strict Sense: If $(V_t,\ W_t)$ are Brownian motions, then $(x_t,\ y_t)$ is a Gaussian process. Then $\hat{x}_t = E[x_t | y_s,\ s \leq t]$ is the "best estimate" in the sense of minimizing $E(x_t - z)^2$ over all $\mathcal{Y}_t$ - measurable, square integrable random variables z, where

$$\mathcal{Y}_t = \sigma\{y_s, s \le t\}.$$

Because of normality, $\hat{x}_t$ is a linear function of $(y_s, s \le t)$.

(b) Wide Sense Formulation: Do not suppose V_t, W_t are normally distributed. Just suppose that the ith coordinates V_t^i, W_t^i are uncorrelated and $E\,V_t^i\,V_s^i = t \wedge s$; $E\,W_t^i\,W_s^i = t \wedge s$, i.e., V^i, W^i are orthogonal increment processes. Now look for the best linear estimate of x_t given $(y_s, s \le t)$. This will coincide with $E(x_t|\mathcal{Y}_t)$ in strict sense case.

Calculating $\hat{x}_t$ is a Hilbert space projection problem. The random variables we consider belong to $L_2^o(\Omega, F, P)$ which is a Hilbert space with inner product $(X,Y) = E\,X\,Y$, where o denotes the elements are of zero mean. For any process, say y_t, define $\mathcal{H}^y = \mathcal{L}(y_t, t \ge o)$, the linear span of y_t; this is a linear subspace. Then if $\hat{z}$ denotes the projection of z onto $\mathcal{H}^y$, then

$$\| z - \hat{z} \| = \min_{U \in \mathcal{H}^y} \| z - U \|.$$

Let $\hat{x}_t$ be projection of x_t onto $\mathcal{H}_t^y = \mathcal{L}(y_s, s \le t)$. Then the "Innovations process" ν_t is defined by

$$d\nu_t = dy_t - H\,\hat{x}_t\,dt. \tag{3}$$

The Innovations process ν_t has the following properties:

(i) ν_t is an orthogonal increments process.

(ii) $\mathcal{H}_t^y = \mathcal{H}_t^\nu$.

(iii) $\mathcal{H}_t^\nu = \{\int_o^t g(s)d\nu_s,\ g \in L_2[o,t]\}$.

Then $\hat{x}_t$ satisfies the linear equation

$$\begin{aligned} d\hat{x}_t &= A\,\hat{x}_t\,dt + P(t)\,H'\,d\nu_t \\ \hat{x}_o &= 0 \end{aligned} \tag{4}$$

where the error covariance $P(t) = E(x_t - \hat{x}_t)(x_t - \hat{x}_t)'$, (' denotes the transpose).

$P(t)$ satisfies the "Riccati equation"

$$\frac{d}{dt} P(t) = AP(t) + P(t)A' + CC' - P'(t)HH'P(t)$$

$$P(o) = P_o = \text{Cov}\,(x_o).$$

The above equation (4) is the <u>Kalman Filter</u>.

<u>Derivation of Kalman Filter equation</u>: From properties (ii), (iii) we know

$$\hat{x}_t = \int_o^t g(t,s)\,d\nu_s$$

for some g such that

$$\int_o^t g^2(t,s)\,ds < \infty.$$

Now using projection, $x_t - \hat{x}_t \perp \nu_s$, $s \leq t$. So

$$\begin{aligned} Ex_t\,\nu_s' &= E\,\hat{x}_t\,\nu_s' \\ &= E(\int_o^t g(t,u)d\nu_u)\nu_s' \\ &= \int_o^t g(t,u)\,du. \end{aligned}$$

Hence

$$g(t, s) = \frac{d}{ds} E\, x_t\, \nu_s'.$$

Write innovations process as

$$d\nu_t = H\,\tilde{x}_t\, dt + dW_t \quad \text{where} \quad \tilde{x}_t = x_t - \hat{x}_t.$$

$$E\, x_t\, \nu_s' = \int_0^s E(x_t\, \tilde{x}_u')H'\, du.$$

Now

$$x_t = \Phi(t, u)\, x_u + \int_u^t \Phi(t, r)\, C\, dV_r$$

where Φ is the transition matrix of A. So

$$E\, x_t\, \nu_s' = \int_0^s \Phi(t,u)\, E(\tilde{x}_u\, \tilde{x}_u')\, H'\, du$$

$$= \int_0^s \Phi(t,u)\, P(u)\, H'\, du$$

$$g(t, s) = \Phi(t, s)\, P(s)\, H'.$$

So

$$\hat{x}_t = \int_0^t \Phi(t, s)\, P(s)\, H'\, d\,\nu_s.$$

But this is the unique solution of (4).

Important Points: (1) It is a recursive estimator.

(2) In the strict sense version $\hat{x}_t$ is a sufficient statistic for the conditional distribution of x_t given $(y_s,\ s \leq t)$, since this distribution is $N(\hat{x}_t, P(t))$ and $P(t)$ is nonrandom.

<u>EXERCISE 1.1</u>: (Constant Signal). Let $x_t = \theta$ with $E(\theta) = 0$, $\operatorname{Var}(\theta) = \sigma^2$ and

$$dy_t = \theta \, dt + dW_t$$

with θ independent of W_t. Show directly by projection that

$$\hat{\theta}_t = \frac{1}{t + \frac{1}{\sigma^2}} y_t .$$

Now show that the Kalman filter gives the same result.

Nonlinear Filtering

Suppose "signal" x_t is a Markov process and "observations" y_t is given by

$$dy_t = h(x_t) \, dt + dW_t,$$

generally h is a bounded measurable function (extra smoothness condition will be added later). Assume that for each t, x_t and $(W_u - W_v)$, $u, v \geq t$ are independent, which allows for the "feedback" case. Our objective is to calculate in recursive form the "estimates" of x_t. To do this, it is necessary to compute the conditional distribution of x_t given

$$\mathcal{Y}_t = \sigma\{y_s, s \leq t\}.$$

The Innovations Approach to Nonlinear Filtering

This approach was originally suggested by Kailath for the linear case and by Kailath and Frost for nonlinear filtering. The definitive formulation of the filtering problem from the innovations standpoint was given by Fujiskki, Kallianpur and Kunita [18].

Innovations Processes: Consider process y_t satisfying

$$dy_t = z_t\, dt + dW_t,\ t \in [o,T] \tag{5}$$

where W_t is Brownian motion and assume

$$E \int_o^T z_s^2\, ds < \infty \tag{6}$$

and the "feedback" condition is satisfied. Let

$$\hat{z}_t = E[z_t | \mathcal{Y}_t].$$

More precisely $\hat{z}_t$ is the "predictable projection" of z_t onto $\mathcal{Y}_t$. The innovations process is then

$$d\nu_t = dy_t - \hat{z}_t\, dt. \tag{7}$$

NOTE (i): ν_t is a Brownian motion w.r.t. $\mathcal{Y}_t$, i.e., ν_t is a $\mathcal{Y}_t$ martingale and $< \nu >_t = t$. If $F_t^\nu = \sigma\{\nu_s, s \leq t\}$, the question is whether $F_t^\nu = F_t^y$. It has been shown that in general, this is not true. But if (i) holds and (z_t), (W_t) are independent, then Allinger-Mitter proved that $F_t^\nu = F_t^y$.

NOTE (ii): All $\mathcal{Y}_t$ - martingales are stochastic integrals w.r.t. (ν_t), i.e. if M_t is a $\mathcal{Y}_t$ - martingale, then there is a g such that

$$\int_o^T g_s^2\, ds < \infty \text{ a.s.}$$

and

$$M_t = \int_o^t g_s\, d\nu_s.$$

This is true even if $F_t^\nu \neq F_t^y$, but note that (g_s) is adapted to F_t^y, not necessarily to F_t^ν.

A GENERAL FILTERING FORMULA: Take an F_t - martingale n_t, process (α_t) satisfying

$$E \int_0^T |\alpha_s|^2 \, ds < \infty$$

and F_o measurable random variable ξ_o with $E\,\xi_o^2 < \infty$. Now define an F_t semi-martingale ξ_t by

$$\xi_t = \xi_o + \int_0^t \alpha_s \, ds + n_t. \tag{8}$$

Since $< W, W >_t = t$, we have

$$< n, W >_t = \int_0^t \beta_s \, ds$$

for some β_t and for any martingale n_t. Let

$$\hat{\xi}_t = E[\xi_t | Y_t].$$

Then $\hat{\xi}_t$ satisfies the following stochastic differential equation

$$\hat{\xi}_t = \hat{\xi}_o + \int_0^t \hat{\alpha}_s \, ds + \int_0^t [\widehat{\xi_s z_s} + \hat{\xi}_s \hat{z}_s + \hat{\beta}_s] d\nu_s. \tag{9}$$

PROOF: Define

$$\mu_t = \hat{\xi}_t - \hat{\xi}_o - \int_0^t \hat{\alpha}_s \, ds$$

Then μ_t is a Y_t - martingale. So there is some integrable function η such that

$$\mu_t = \int_0^t \eta_s \, d\nu_s. \tag{10}$$

Now we will identify the form of η_t, using ideas of Wong [28]. Using (5) and (8) and Ito formula,

$$\xi_t y_t = \xi_o y_o + \int_o^t \xi_s (z_s ds + dW_s) + \int_o^t y_s(\alpha_s ds + dn_s)$$

Now calculate $\hat{\xi}_t\, y_t$ using (7) and (10),

$$\hat{\xi}_t y_t = \hat{\xi}_o y_o + \int_o^t \hat{\xi}_s (\hat{z}_s ds + d\nu_s) + \int_o^t y_s(\hat{\alpha}_s ds + \eta_s d\nu_s)$$

$$+ \int_o^t \eta_s\, ds.$$

Now for $t \geq s$,

$$E[\xi_t y_t - \hat{\xi}_t y_t | \mathcal{Y}_s] = 0.$$

So

$$E[\int_s^t (\widehat{\xi_u z_u} - \hat{\xi}_u \hat{z}_u + \hat{\beta}_u - \eta_u) du | \mathcal{Y}_s] = 0.$$

Let

$$V(u) = \widehat{\xi_u z_u} - \hat{\xi}_u \hat{z}_u + \hat{\beta}_u - \eta_u.$$

Then $V(u)$ is a predictable process and

$$E[\int_s^t V(u) du | F_s] = 0.$$

This is,

$$\int_{A \times]s,t]} V(u) du\, dP = 0 \quad \forall\ s,\ t \geq s,\ A \in F_s.$$

The class of sets $A \times]s,t]$ generates P, the predictable σ - field. Hence $V(u,\omega) = 0$ a.e. $dt * dP$. Hence the result.

Formula (9) is not a recursive equation for $\hat{\xi}_t$. Still we can use it to obtain more explicit results for filtering of Markov processes. Let (x_t) be a Markov process and $A, \mathcal{D}(A)$ be its generator, i.e. for $f \in \mathcal{D}(A)$ then

$$C_t^f = f(x_t) - f(x_s) - \int_s^t Af(x_u)du$$

is a martingale. Suppose

$$< C^f, W >_t = \int_o^t Z f(x_s) ds$$

for some function Zf. Introduce the notation

$$\Pi_t(f) = E[f(x_t)|\mathcal{Y}_t].$$

Now apply (9) with $\xi_t = f(x_t)$; $Af(x_s) = \alpha_s$, $C_t^f = n_t$ and $z_t = h(x_t)$ to get the Fujisaki-Kallianpur -Kunita filtering formula

$$\Pi_t(f) = \Pi_o(f) + \int_o^t \Pi_s(Af)ds + \int_o^t [\Pi_s(Df) - \Pi_s(h)\Pi_s(f)]d\nu_s \quad (11)$$

where

$$Df(x) = Zf(x) + h(x) f(x).$$

If we interpret Π_t as the conditional distribution of x_t given $\mathcal{Y}_t$, so that

$$\Pi_t(f) = \int f(x) \Pi_t(dx) = E[f(x_t)|\mathcal{Y}_t],$$

then (11) is a measure-valued stochastic differential equation, and gives an infinite-dimensional recursive equation for filtering.

EXERCISE 1.2: Derive the Kalman filter from the Fujisaki-Kallianpur-Kunita equation.

The Unnormalized (Zakai) Equations:

Introduce a new probability measure P_o on (Ω,F) with $t \in [o,T]$ by

$$\frac{dP_o}{dP} = \exp\left(-\int_o^T h(x_s)dW_s - \frac{1}{2}\int_o^T h^2(x_s)\,ds\right).$$

Since h is bounded, P_o is a probability measure and (y_t) is a P_o - Brownian motion. Also

$$\begin{aligned} < C_t^f, y >_t &= < C_t^f, w >_t \\ &= \int_o^t Zf(x_s)\,ds. \end{aligned}$$

Note that, in general, C_t^f is a semi-martingale under P_o but $<.,.>$ is invariant under absolutely continuous change of measure. Also if $Z = o$, then x_t has the same distributions under either measure. Let

$$\Lambda_T = \frac{dP}{dP_o} = \exp\left(\int_o^T h(x_s)dy_s - \frac{1}{2}\int_o^T h^2(x_s)ds\right).$$

Let E_o denote the expectation under P_o. Then it can be calculated that under measure P_o,

$$\begin{aligned} \Pi_t(f) &= E[f(x_t)|\mathcal{Y}_t] \\ &= \frac{E_o[f(x_t)\,\Lambda_t|\mathcal{Y}_t]}{E_o[\Lambda_t|\mathcal{Y}_t]} \\ &=: \frac{\sigma_t(f)}{\sigma_t(1)}. \end{aligned}$$

Then $\sigma_t(f)$ is an unnormalized conditional distribution since $\sigma_t(1)$ does not depend on f. To obtain an equation satisfied by σ_t, we need a semi-martingale representation for $\sigma_t(1)$. First we have

$$d\Lambda_t = h(x_t)\Lambda_t \, dy_t \tag{12}$$

i.e. $$\Lambda_t = 1 + \int_0^t h(x_s)\Lambda_s \, dy_s$$

also Λ_t is a (F_t, P_o) martingale. Then as before

$$\hat{\Lambda}_t = E_o[\Lambda_t | \mathcal{Y}_t]$$

is a $\mathcal{Y}_t$ - martingale, so there exists some $\mathcal{Y}_t$ - adapted integrand η_t such that

$$\hat{\Lambda}_t = 1 + \int_0^t \eta_s \, dy_s \, . \tag{13}$$

To identify η_t, we use the same technique as in deriving the FKK equation. Calculate using (12) and Ito's rule,

$$\Lambda_t y_t = \int_0^t \Lambda_s \, dy_s + \int_0^t y_s \Lambda_s h(x_s) dy_s + \int_0^t \Lambda_s h(x_s) \, ds.$$

Calculate using (13) and Ito's rule,

$$\hat{\Lambda}_t y_t = \int_0^t \hat{\Lambda}_s dy_s + \int_0^t y_s \eta_s \, dy_s + \int_0^t \eta_s \, ds.$$

Now

$$E_o[\Lambda_t y_t - \hat{\Lambda}_t y_t | \mathcal{Y}_s] = 0 \quad \text{for} \quad t \geq s,$$

so we get

$$\eta_t = \widehat{\Lambda(t)h(x_t)} := E_o[\Lambda_t h(x_t)|\mathcal{Y}_t].$$

So (13) becomes

$$\hat{\Lambda}_t = 1 + \int_o^t \hat{\Lambda}_s \Pi_s(h)\, dy_s. \tag{14}$$

This has a unique solution

$$\hat{\Lambda}_t = \exp\left(\int_o^t \Pi_s(h)\, dy_s - \frac{1}{2}\int_o^t \Pi_s^2(h)\, ds\right)$$

$$= \sigma_t(1).$$

THEOREM 1.1: $\sigma_t(f)$ satisfies the "Zakai equation"

$$d\sigma_t(f) = \sigma_t(Af)dt + \sigma_t(Df)dy_t \tag{15}$$

$$\sigma_o(f) = \Pi_o(f) = E[f(x_t)].$$

PROOF: Direct calculation using (11), (14) and the fact that

$$\sigma_t(f) = \hat{\Lambda}_t\, \Pi_t(f).$$

COROLLARY 1.1: There is a one-to-one relation between Zakai equation and FKK equation, in that whenever σ_t satisfies Zakai equation, $\sigma_t(f)/\sigma_t(1)$ satisfies (11), and whenever $\Pi_t(f)$ satisfies (11),

$$\Pi_t(f)\exp\left(\int_o^t \Pi_s(h)dy_s - \frac{1}{2}\int_o^t \Pi_s(h)ds\right)$$

satisfies Zakai equation.

The Zakai Equation with Stratonovich Integrals:

Recall

$$\int_0^t u_s \circ dv_s := \int_0^t u_s \, dv_s + \frac{1}{2} < u, v >_t$$

where $\circ$ denotes a Stratonovich Stochastic integral and u and v are continuous semi-martingales. We have to calculate $< \sigma_.(Df), y >_t$. From Zakai equation

$$d\sigma_t(Df) = \sigma_t(ADf)\, dt + \sigma_t(D^2 f) dy_t.$$

So

$$d < \sigma_.(Df), y >_t = \sigma_t(D^2 f) dt.$$

So the Stratonovich version of the Zakai equation is

$$d\sigma_t(f) = \sigma_t(Af)dt + \sigma_t(Df)\circ dy_t - \frac{1}{2} \sigma_t(D^2 f)dt$$

$$= \sigma_t(Lf)\, dt + \sigma_t(Df)\circ dy_t$$

where

$$Lf(x) = Af(x) - \frac{1}{2} D^2 f(x).$$

Application to Diffusion Process:

Consider a process $x_t \in \mathbb{R}^d$ satisfying

$$df(x_t) = X_o f(x_t)dt + X_j f(x_t) \circ dB_t^j \qquad (16)$$

for arbitrary smooth f, where $X_o, \ldots, X_r$ are vector fields on $\mathbb{R}^d$ we suppose that $< B^j, W >_t = \alpha^j t$ for some constants $\alpha^1, \ldots, \alpha^r$

Note that A is the generator of x_t under measure P (not P_o). This is given by

$$Af(x) = X_o f(x) + \frac{1}{2} \sum_j X_j^2 f(x).$$

PROOF: Rewrite (16) in Ito form. Replace f by $X_k f$ in (16).

$$dX_k f(x_t) = X_o X_k f(x_t)dt + X_j X_k f(x_t) \circ dB_t^j.$$

Then

$$d < X_k f, B^k >_t = X_k^2 f(x_t)dt.$$

Then Ito version of (16) is

$$df(x_t) = (X_o + \frac{1}{2} \sum X_j^2)f(x_t)dt + X_j f(x_t)dB_t^j.$$

So

$$A = X_o + \frac{1}{2} \sum X_j^2$$

and*

$$C_t^f = \int_o^t X_j f(x_s)dB_s^j.$$

PROPOSITION 1.1: For Z given by

$$< C^f, W >_t = \int Zf(x_s)ds.$$

with $Zf = \alpha^j X_j$, Z is a vector field.

PROOF:
$$\begin{aligned} d < C^f, W > &= d < \int X_j f dB^j, W > \\ &= X_j f d < B^j, W > \\ &= \alpha^j X_j f(x_t) dt. \end{aligned}$$

---

* We sometimes use the convention of implied summation over repeated indices.

So

$$D = Z + h$$
$$= \alpha^i X_i + h.$$

<u>PROPOSITION 1.2</u>: <u>There exist vector fields</u> $Y_o, Y_1, \ldots, Y_r$ <u>such that</u>

$$A - \frac{1}{2} D^2 = \frac{1}{2} \sum_j Y_j^2 + Y_o - \frac{1}{2} Dh.$$

<u>PROOF</u>:

$$D^2 f = (\alpha^i X_i + h)(\alpha^j X_j f + hf)$$
$$= \alpha^i \alpha^j X_i X_j f + \alpha^i X_i (hf) + \alpha^j h X_j f + h^2 f$$
$$= \alpha^i \alpha^j X_i X_j f + hZf + Dhf.$$

Let $\alpha = (\alpha^1, \alpha^2, \ldots \alpha^r)'$ and suppose $I - \alpha\alpha'$ is nonnegative definite. Write $(I - \alpha\alpha') = \Delta\Delta'$ and let $X = (X_1, \ldots, X_r)'$.

$$X' \Delta\Delta' X = \sum X_i^2 - \alpha^i \alpha^j X_i X_j.$$

So define $Y = \Delta' X$. Then Y_i's are vector fields and

$$A - \frac{1}{2} D^2 = \frac{1}{2} \sum_i Y_i^2 - hZ + X_o - \frac{1}{2} Dh$$
$$= \frac{1}{2} \sum_i Y_i^2 + Y_o - \frac{1}{2} Dh$$

where $Y_o = X_o - hZ$. It remains to check that $I - \alpha\alpha' \geq o$. Take $\xi \in \mathbb{R}^r$ with $|\xi| = 1$. Then

$$\xi'(I - \alpha\alpha')\xi = 1 - (\alpha'\xi)^2$$
$$\geq 1 - |\alpha|^2$$
$$\geq 0.$$

Since $|\alpha|^2 \leq 1$, for

$$< \sum_i \alpha^i B^i, W >_t = \sum_i \alpha_i < B^i, W >_t$$

$$= |\alpha|^2 t$$

So $\alpha^2 t = E\, U\, W_t$, where $U = \sum_i \alpha^i B^i$.

$$|\alpha|^4 t^2 \leq E\, U^2\, t = |\alpha|^2 t^2.$$

So we have Zakai equation in Ito's form

$$d\sigma_t(f) = \sigma_t(Af)dt + \sigma_t(Df)dy_t$$

with vector fields $X_o, X_1, \ldots X_r$ and in Stratonovich form

$$d\sigma_t(f) = \sigma_t(Lf) + \sigma_t(Df) \circ dy_t$$

with vector fields $Y_o, Y_1, \ldots Y_r$ plus a "0th-order" term. Now we investigate what sort of process is x_t under the measure P_o.

PROPOSITION 1.3: Under P_o, x_t satisfies the equation

$$df(x_t) = Y_o f(x_t)dt + Zf(x_t) \circ dy_t + Y_j f(x_t) \circ db_t^j \quad (17)$$

where $b^1, \ldots b^r$ are independent standard Brownian motions independent of y_t.

PROOF: Recall the Girsanov transformation. Under P, $B^1, \ldots B^r$ are independent and $< B^j, W >_t = \alpha^j t$. Now

$$\frac{dP_o}{dP} = \exp\left(M_t - \frac{1}{2} < M, M >_t\right)$$

where M is a P - martingale. Under P_o, $B_t^j - < B^j, M >_t$

is a martingale and hence a Brownian motion. Here

$$M_t = -\int_0^t h(x_s)dW_s;$$

$$d < B^j, M > = -\alpha^j h(x_t)dt.$$

So under P_o,

$$dV_t^j = dB_t^j + \alpha^j\, h(x_t)\, dt$$

are independent Brownian motions, but V^j is not independent of y_t, in fact $< V^j, y >_t = \alpha^j\, t$.

Now define $\tilde{b}_t^j = V_t^j - \alpha_j y_t$, then $< \tilde{b}^j, y > = 0$, and this implies $\tilde{b}^j$, y are independent. But the $\tilde{b}^j$ are now not independent. In fact,

$$< b^j, b^k >_t = \begin{cases} -\alpha^j\alpha^k t & \text{for } k \neq j \\ 1 - (\alpha^j)^2 t & \text{for } k = j \end{cases}$$

So

$$< \tilde{b}^j, \tilde{b}^k >_t = [< \tilde{b}^j, \tilde{b}^k >_t]$$

$$= (I - \alpha\alpha')t.$$

Let $(I - \alpha\alpha') = \Delta\Delta'$ as before and define $b_t = \Delta'^{-1}\tilde{b}_t$. Then $\tilde{b}_t = \Delta' b_t$ and $< b >_t = It$. So

$$df(x_t) = X_o f(x_t)dt + X_j f(x_t) \circ dB^i$$

$$= X_o f(x_t)dt + X_j f(x_t) \circ (-\alpha^j h(x_t)dt + \alpha^j dy_t + d\tilde{b}_t^j)$$

$$= (X_o f(x_t) - hZf(x_t))dt + Zf(x_t) \circ dy_t + Y_j f(x_t) \circ db_t^j$$

$$= Y_o f(x_t)dt + Zf(x_t) \circ dy_t + Y_j f(x_t) \circ db_t^j$$

where $Y_o f = X_o f - hZf$ and $Y = \Delta' X$.

The so called Kallianpur-Striebel Formula gives the solution σ_t of the Zakai equation as a function space integral in the following form, where x_t is functional of $(y, b^1, \ldots b^r)$.

$$\sigma_t(f) = E_o[f(x_t)\exp(\int_o^t h(x_s)dy_s - \frac{1}{2}\int_o^t h^2(x_s)ds) | \mathcal{Y}_t]$$

i.e. as a function of y, we have

$$\sigma_t(f)(y) = \int_{C^r[o,t]} [f(x_t)\exp(\int_o^t h(x_s)dy_s$$

$$- \frac{1}{2}\int_o^t h^2(x_s)ds] \mu_w(db^1)\ldots\mu_w(db^r)$$

where $\mu_w(db)$ is the Wiener measure on $C[o,T]$.

§2. PATHWISE SOLUTIONS OF DIFFERENTIAL EQUATIONS

Consider the Doss-Sussman construction for the equation

$$\dot{x} = b(x) + g(x)\dot{w}$$
$$x(o) = x \qquad (1)$$

where $w \in C^1(\mathbb{R}_+)$. Let $\phi(t,x)$ be the "flow" of g, i.e.

$$\frac{\partial}{\partial t}\phi(t,x) = g(\phi(t,x))$$
$$\phi(o,x) = x.$$

If $b = o$, then it is immediate that the solution of (1) is

$$x_t = \phi(w(t), x).$$

If $b \neq o$, then the solution of (1) is of the form

$$x_t = \phi(w(t), \eta(t)) \qquad (2)$$

where $\eta(t)$ satisfies some ODE. With $x(t)$ defined by (2),

$$\dot{x}(t) = g(x(t))\dot{w}(t) + \phi_x(w(t),\eta(t))\dot{\eta}(t)$$

and we require that

$$\phi_x(w(t),\eta(t))\dot{\eta}(t) = b(\phi(w(t),\eta(t))).$$

So $x(t)$ satisfies (1) if $\eta(t)$ satisfies

$$\dot{\eta}(t) = (\phi_x(w(t),\eta(t)))^{-1}\, b(\phi(w(t),\eta(t)))$$

$$\eta(o) = x.$$

<u>Coordinate-free form:</u> Let

$$X_o\, f(x) = b(x)\,\frac{df}{dx}$$
$$X_1\, f(x) = g(x)\,\frac{df}{dx}$$

$$\xi_t\,(x) = \phi(w(t),x).$$

Define

$$(\xi^{-1}_{t*}\, X_o)f(x) = X_o(f \circ \xi_t^{-1})(\xi_t(x)).$$

Then the equation for $\eta(t)$ can be expressed as

$$\frac{d}{dt} f(\eta_t) = (\xi_{t*}^{-1} X_o) f(\eta_t)$$

$$\eta_o = x ,$$

for

$$(\xi_{t*}^{-1} X_o) f(\eta_t) = b(\xi_t(x)) \frac{d}{dx} f(\xi_t^{-1}(x))|_{\xi_t(x)}$$

$$= b(\xi_t(x)) \frac{d}{dx} f(x) \frac{d}{dx} (\xi_t^{-1}(x))|_{\xi_t(x)}$$

$$= b(\xi_t(x)) \frac{d}{dx} f(x) (\xi_x(x))^{-1}. \qquad (*)$$

Since $\xi^{-1}(\xi(x)) = x$ and so

$$\frac{d}{dx} (\xi^{-1}(\xi(x))) \frac{d}{dx} \xi(x) = 1.$$

When $x \in \mathbb{R}^d$, then

$$X_o f(x) = \sum_{i=1}^{d} b^i(x) \frac{\partial f(x)}{\partial x_i}$$

$$X_1 f(x) = \sum_{i=1}^{d} g^i(x) \frac{\partial f(x)}{\partial x_i} .$$

Then (*) is of the form

$$(\xi_{t*}^{-1} X_o) f(\eta_t) = \sum_{i=1}^{d} b^i(x) \frac{\partial}{\partial x_i} \{f \circ \xi_t^{-1}\}|_{\xi_t(x)}$$

$$= \sum_{j=1}^{d} \sum_{i=1}^{d} \frac{\partial f(x)}{\partial x_i} \frac{\partial (\xi_t^{-1})^j}{\partial x_i} (x).$$

So the same results apply for $x_t \in \mathbb{R}^d$, but generally <u>not</u> for more than one "input", i.e. for vector $w(t)$.

<u>Interpretation</u>: x_t defined by (2) makes sense for any $w(.) \in C(\mathbb{R}_+)$. In particular, if $w(t)$ is a sample path of

Brownian motion, then what equation does x_t satisfy?

Answer: the Stratonovich equation

$$dx_t = b(x_t)dt + g(x_t) \circ dw_t \tag{3}$$

<u>EXERCISE 2.1</u>: Expand x_t given by (2) using Ito's calculus and show that it satisfies (3).

The following examples show that the pathwise solution idea cannot generally be extended to "multi-input" equations.

<u>EXAMPLE 2.1</u>: Let

$$\dot{x} = g^1(x)\,\dot{w}^1 + g^2(x)\,\dot{w}^2$$

$$x(o) = x.$$

The solution should be of the form

$$x_t = h(w_t^1 , w_t^2) . \tag{4}$$

Then with $h_1(w^1,w^2) = \dfrac{\partial}{\partial w^1} h(w^1,w^2)$ etc. , we have

$$\dot{x}_t = h_1\,\dot{w}^1 + h_2\,\dot{w}^2$$

$$h_1(w_t^1, w_t^2) = g^1 \circ h(w_t^1, w_t^2)$$

$$h_2(w_t^1, w_t^2) = g^2 \circ h(w_t^1, w_t^2)$$

and

$$h_{12}(w_t^1, w_t^2) = g_x^1 \circ h.h_2 = (g_x^1 \circ h)(g^2 \circ h)$$

$$h_{21}(w_t^1, w_t^2) = (g_x^2 \circ h)(g^1 \circ h).$$

So we must have

$$g^1 g^2_x = g^2 g^1_x .$$

Define the Lie bracket $[X_1, X_2] = X_1 X_2 - X_2 X_1$. Now

$$X_1 X_2 f = g^1 \frac{d}{dx} (g^2 \frac{df}{dx})$$

$$= g^1 g^2_x f_x + g^1 g^2 f_{xx} .$$

Therefore

$$[X_1, X_2] f = (g^1 g^2_x - g^2 g^1_x) f_x .$$

So a necessary condition for (4) to hold is that

$$[X_1, X_2] = 0$$

i.e.,

$$X_1 X_2 = X_2 X_1 .$$

<u>EXERCISE 2.2:</u> Consider

$$\dot{X} = \sum_{i=1}^{n} g^i(x) \dot{w}^i .$$

Let $\phi^i(t,x)$ be the flow of g^i and $\xi^i_t(x) = \phi^i(w^i_t, x)$.

Then show that

$$x_t = \xi^1_t \circ \xi^2_t \circ \dots \circ \xi^n_t$$

if $[X^i, X^j] = 0 \quad \forall \; i,j$.

With one input, $\| w^n - w \| \to o$ implies $x^n_t \to x_t$, where $\| . \|$ is the sup norm. But with inputs w^1, w^2, the solution map generally is not continuous.

<u>EXAMPLE 2.2</u> (Sussmann [23]): Let $t \in [o,1]$ and

$$\dot{x}^n = Ax^n\dot{w}^{1,n} + Bx^n\dot{w}^{2,n}$$

$$x(o) = x_o$$

where A, B are $n \times n$ matrices with

$$[A, B] = AB - BA \neq 0.$$

Partition $[0,1]$ into n equal intervals $I_j^n = [\frac{j-1}{n}, \frac{j}{n}]$, $j = 1,2,\ldots,n$. Partition each I_j^n into four equal intervals $I_{j,i}^n$, $i = 1,2,3,4$. Define $w^{1,n}$ to be equal to $4n^{1/2}$ for $t \in I_{j,1}^n$ to $-4n^{1/2}$ for $I_{j,2}^n$, and to zero for all other t. Similarly, let $\dot{w}^{2,n}$ be equal to $4n^{1/2}$ for $t \in I_{j,2}^n$ to $-4n^{1/2}$ for $t \in I_{j,4}^n$, and to zero for all other t. Then

$$w^{i,n}(t) = \int_o^t \dot{w}^{i,n}(s)\, ds, \quad i.= 1,2.$$

Clearly $\dot{w}^{i,n}$ converges to zero uniformly as $n \to \infty$, $i = 1,2$. Let $s = n^{-1/2}$, then

$$x^n(1/n) = e^{-Bs}\, e^{-As}\, e^{Bs}\, e^{As}\, x_o$$

$$= e^{\tau}\, x_o.$$

We use the Baker-Campbell-Hausdorff formula $e^A e^B = e^C$ where

$$C = A + B + \frac{1}{2}[A,B] + \frac{1}{12}\{[[B,A],A]+[[B,A],B]\} + \ldots$$

we get

$$\tau = [B,A]\frac{1}{n} + o(1/n) \cdot$$

So

$$x^n(1) = e^{n\tau}\, x_o$$

$$= e^{([B,A]+\, o\,(1/n))}\, x_o$$

Hence

$$\lim_{n \to \infty} x_t^n = e^{t[B,A]}\, x_o.$$

3. PATHWISE SOLUTION OF THE FILTERING EQUATION

Consider the equation

$$df(x_t) = Y_o f(x_t)dt + Zf(x_t) \circ dy_t + Y_i f(x_t) \circ db_t^i.$$

To express this pathwise in (y_t), let $\phi(t,x)$ be the integral curve of Z and $\xi_t(x) = \phi(y(t), x)$. Define η_t as follows:

$$df(\eta_t) = (\xi_{t*}^{-1}\, Y_o)f(\eta_t)dt + (\xi_{t*}^{-1}\, Y_j)f(\eta_t) \circ db_t^j$$

$$\eta_o = x.$$

Then

$$x_t = \xi_t \circ \eta_t$$

$$= \phi(y(t), \eta(t)).$$

The generator of η is

$$A_t^* = \xi_{t*}^{-1}\, Y_o + \frac{1}{2} \sum_j (\xi_{t*}^{-1}\, Y_j)^2.$$

The Kallianpur-Striebel Formula: Recall

$$\sigma_t(f) = E^{(b)}[f(x_t)\exp(\int_0^t h(x_s)dy_s - \frac{1}{2}\int_0^t h^2(x_s)ds)]$$

$$= E^{(b)}[f(x_t)\exp(\int_0^t h(x_s) \circ dy_s - \frac{1}{2}\int_0^t Dh(x_s)ds)]$$

where $D = Z + h$.

NOTATION: For any diffeomorphism $\psi: M \to M$, $\psi^*: C^\infty(M) \to C^\infty(M)$ is given by

$$\psi^* f(x) = f \circ \psi(x) = f(\psi(x)).$$

So

$$f(x_t) = \xi_t^* f(\eta_t),$$

and

$$\sigma_t(f) = E^{(b)}[\xi_t^* f(\eta_t)\exp(\int_0^t \xi_s^* h(\eta_s) \circ dy_s -$$

$$- \frac{1}{2}\int_0^t \xi_s^* Dh(\eta_s)ds\ ]$$

The next step is to remove "$\circ dy$". Define

$$H(t,x) = \int_0^t \phi_s^* h(x)\, ds.$$

Calculate $H(y_t, \eta_t)$ using Stratonovich calculus

$$dH(y_t,\eta_t) = \xi_t^* h(\eta_t)\circ dy_t + \xi_{t*}^{-1} Y_0 Hy_s(\eta_s)ds$$

$$+ \xi_{t*}^{-1} Y_i\ Hy_s(\eta_s)\circ\, db_s^i.$$

NOTATION: $g_s(x) = H(y(s), x)$

$$Y_j^* = \xi_{t*}^{-1}\, Y_j$$

$$B_s f(x) = \phi_s^* f(x) \exp\left(\int_o^s \phi_u^* h(x)du\right).$$

Finally, we get

$$\sigma_t(f) = E^{(b)}[By_{(t)} f(\eta_t) \alpha_t^o(y)]$$

where the multiplicative functional α_t^s is given by

$$\alpha_t^s(y) = \exp[\int_s^t Y_j^* g_u(\eta_u)db_u^j - \frac{1}{2}\int_s^t (Y_j^*)^2 g_u(\eta_u)du$$

$$- \int_s^t Y_o^* g_u(\eta_u)du - \frac{1}{2}\int_s^t \xi_u^* Dh(\eta_u)du]$$

So $\sigma_t(f)$ is now pathwise in y with $\sigma_t(f)\colon C[o,t] \to \mathbb{R}$ and $(\sigma_t(f)/\sigma_t(1))$ is a version of $E[f(x_t)|\mathcal{Y}_t]$. Now we want to compute $\sigma_t(f)$ recursively.

(a) Multiplicative Functional Approach:

Let (x_t) be a Markov process with extended generator $(A, \mathcal{D}(A))$. The associated semigroup on $\mathcal{B}(E)$ is

$$T_{s,t} f(x) = E_{s,x} [f(x_t)]$$

$\alpha_t^s (s \leq t)$ is a multiplicative functional (m.f.) of (x_t) if α_t^s is $\sigma\{x_u, s \leq u \leq t\}$ - measurable and for $r \leq s \leq t$,

$$\alpha_t^r = \alpha_s^r \alpha_t^s.$$

Corresponding to α_t^s there is a semigroup defined by

$$T_{s,t}^\alpha f(x) = E_{s,x} [f(x_t) \alpha_t^s].$$

In particular,

$$T_{s,t}^\alpha 1 = E_{s,x} [\alpha_t^s].$$

It is a Markov (or Sub-Markov) semigroup when

$$E_{s,x}\,[\alpha_t^s] = 1 \ (\leq 1).$$

If (x_t) is a homogeneous Markov process, α_t^s is a homogeneous m.f. if

$$\alpha_t^s = \alpha_{t+r}^{s+r} \circ \theta_r$$

where

$$\theta_r\, x_t = x_{t+r}.$$

Then

$$\alpha_t^s = \alpha_{t-s}^o \circ \theta_{-s}.$$

So denoting $\alpha_t = \alpha_t^o$, the m.f. property is

$$\alpha_{t+s} = \alpha_t \cdot \alpha_s \circ \theta_t.$$

Now we want to find the generator of

$$T_t^\alpha\, f(x) = E_x(f(x_t)\ \alpha_t).$$

Suppose for the moment that $\alpha_t \leq 1$, $\forall\, t$. Then α_t is monotone decreasing. In this case, α_t corresponds to "killing" at rate $(-d\alpha_t/\alpha_t)$. It is possible to construct an "α - subprocess" which is a Markov process x_t such that

$$E_x[f(x_t^\alpha)] = T_t^\alpha\, f(x).$$

See Blumenthal and Getoor [1]. Define the extended genrator of T^α to be the extended generator of x_t^α, i.e.,

$$f(x_t^\alpha) - f(x_o^\alpha) - \int_o^t A^\alpha f(x_s^\alpha)ds$$

is a local Martingale if $f \in \mathcal{D}(A^\alpha)$. This says (excluding stopping)

$$E[f(x_t^\alpha) - f(x_s^\alpha) - \int_s^t A^\alpha f(x_u^\alpha)du|F_s] = 0$$

or

$$E_{x_s}[\alpha_{t-s}\, f(x_t)] - f(x_s) - E_{x_s} \int_s^t \alpha_{u-s} A^\alpha f(x_u)du = 0.$$

So equivalently, $f \in \mathcal{D}(A^\alpha)$ if

$$\alpha_t\, f(x_t) - f(x) - \int_o^t \alpha_s A^\alpha f(x_s)ds$$

is a local Martingale (P_x) for every x.

This characterizes A^α even when the condition $\alpha_t \le 1$ is not satisfied, so we adopt it as our definition.

EXAMPLE 3.1: Let $\gamma_t = \exp\,(-\int_o^t V(x_s)ds)$ where $V \in \mathcal{B}(E)$. Take $f \in \mathcal{D}(A)$ and compute

$$d(\gamma_t\, f(x_t)) = \gamma_t\, A\, f(x_t)\, dt + \gamma_t\, d\, M_t\, f - V(x_t)\, f(x_t)\, \gamma_t\, dt.$$

$$\gamma_t f(x_t) - f(x) = \int_o^t \gamma_s[Af(x_s) - V(x_s)f(x_s)]\, ds + \int_o^t \gamma_s\, dM_s f.$$

So

$$A^\gamma f(x) = Af(x) - V(x)f(x).$$

EXAMPLE 3.2: Let

$$\beta_t = \frac{a(x_t)}{a(x_o)}$$

where $a \in \mathcal{D}(A)$ and $a(x) > o \ \forall x$. Then

$$T_t^\beta g(x) = \frac{1}{a(x)} T_t(af)(x).$$

EXERCISE 3.1: Show that

$$A^\beta f(x) = \frac{1}{a(x)} A(af)(x).$$

Now suppose x_t satisfies

$$df(x_t) = X_o f(x_t)dt + X_j f(x_t) \circ dw_t^j.$$

Take $g \in C_b^\infty(E)$ and define

$$\delta_t = \exp(-\int_o^t X_j g(x_u)dw_u^j - \frac{1}{2}\int_o^t \sum_j (X_j g(x_u))^2 du). \qquad (1)$$

If we define

$$\frac{dP_x^\delta}{dP_x} = \delta_t,$$

then

$$d\tilde{w}_t^j = dw_t^j + X_j g(x_t)dt$$

is a P_x^δ - Brownian motion. Thus

$$df(x_t) = (X_o f(x_t) - \sum_j X_j g(x_t) X_j f(x_t))dt + X_j f(x_t) \circ d\tilde{w}_t^j.$$

Now δ_t is a m.f. of x_t (as will be verified below) and

$$E_x[f(x_t)\delta_t] = E_x^\delta[f(x_t)].$$

So

$$A^{\delta}f(x) = (X_o - \sum_j X_j g(x_t) X_j)f + \frac{1}{2}\sum_j X_j^2 f.$$

The three examples here are related by

$$\int_o^t X_j\, g(x_s) dw_s^j = g(x_t) - g(x) - \int_o^t Ag(x_s)ds.$$

Using this in (1), we see that δ_t factors

$$\delta_t = \beta_t\, \gamma_t$$

with

$$V(x) = -\, Ag(x) + \frac{1}{2}\sum_j (X_j\, g(x))^2$$

$$a(x) = e^{-g(x)}.$$

So

$$A^{\delta}f(x) = e^g\, A(e^{-g}f) - (Ag - \frac{1}{2}\sum_j (X_j g)^2)f(x).$$

So

$$e^g A(e^{-g}f) = Af - \sum (X_j g)\, X_j f - (Ag + \frac{1}{2}\sum_j (X_j g)^2)f.$$

<u>EXERCISE 3.2</u>: Verify that this result is correct by direct calculation of $e^g A(e^{-g}f)$.

We have the unnormalized solution of filtering problem as

$$\sigma_t(f) = E[B_{y_t}\; f(\eta_t)\; \alpha_t^o(y)]$$

for $y \in C[o,T]$. Here η_t is a diffusion, $\alpha_t^o(y)$ is a m.f. of η_t. Now

$$\alpha_t^s(y) = \exp(-\int_s^t Y_j^* g(\eta_u)db_u^j - \frac{1}{2}\int_s^t \sum_j (Y_j^* g_u(\eta_u))^2 du)$$

$$\times \exp(\frac{1}{2}\int_s^t \sum_j (Y_j^* g_u(\eta_u))^2 du - \frac{1}{2}\int_s^t (Y_j^*)^2 g_u(\eta_u)du$$

$$- \int_s^t Y_o^* g_u(\eta_u)du - \frac{1}{2}\int_s^t \xi_u^* Dh(\eta_u)du).$$

This factors α_t^s into product of a "Girsanov" m.f. and a "Feynmann-Kac" m.f. Hence the corresponding generator is

$$A_t^y f = A_t^* f - \sum_j Y_j^* \, g_t \, Y_j^* f$$

$$+ [\frac{1}{2}\sum_j (Y_j^* g_t)^2 - A^* g_t - \frac{1}{2}\xi_t^* Dh]f.$$

<u>PROPOSITION 3.1</u>: $A_t^y f = B_{y_t}(A - \frac{1}{2}D^2)B_{y_t}^{-1}$.

<u>PROOF</u>: This can be verified by a straightforward but somewhat lengthy calculation, using the expansion for $e^g A e^{-g}$ obtained previously, once one has obtained an expression for B_t^{-1}. Recall that B_t is defined by

$$B_t f(x) = f(\xi(t,x)) \exp \int_o^t h(\xi(u,x))du.$$

It is a group of operators with generator $D = Z + h$. The inverse B_t^{-1} is given as follows. Let $g(x) = B_t f(x)$, then

$$f(x) = B_t^{-1} g(x)$$

$$= g(\xi(-t,x)) \exp (-\int_o^t h(\xi(u,\xi(-t,x))du)$$

$$= g(\xi^{-1}(t,x)) \exp (-\int_o^t h(\xi^{-1}(s,x))ds).$$

EXAMPLE 3.3: (Independent "signal" and "noise").

Take $Z = 0$, then $\xi(t,x) = x$ and

$$A_t^y f(x) = e^{h(x)y(t)}(A - \frac{1}{2} h^2)(e^{-y(t)h(.)}f(.))(x)$$

$$= e^{hy(t)} A e^{-hy(t)} f - \frac{1}{2} h^2 f.$$

It is easy to see that this must be the right formula. The calculations have been carried out for arbitrary $y \in C[o,T]$ but A_t^y depends only on $y(t)$. So $A_t^y = A_t^{\bar{y}}$ where $\bar{y}(s) \equiv y(t)$ (t fixed). Now

$$\sigma_t(f)(\bar{y}) = E[f(x_t)e^{h(x_t)\bar{y}_t} \exp(-\int_o^t \bar{y}(s)dh(x_s) - \frac{1}{2}\int_o^t h^2(x_s)ds)]$$

$$= E[f(x_t)e^{h(x_t)\bar{y}_t} \exp(-\bar{y}(t)h(x_s) + \bar{y}(t)h(x_o) - \frac{1}{2}\int_o^t h^2(x_s)ds)]$$

$$= E[f(x_t)e^{h(x_t)y_t} \frac{\exp(-y(t)h(x_t))}{\exp(-y(t)h(x_o))} \cdot \exp(-\frac{1}{2}\int_o^t h^2(x_s)ds)].$$

So we have separated into two functionals and the result follows.

<u>Direct Solution of Zakai Equation</u>: We will consider a slight generalization from the original Zakai equation. Define

$$L = \frac{1}{2} \sum_j Y_j^2 + Y_o + h_o$$

$$D = Z + h$$

where Y_i, Z are smooth vector fields; h, h_o are C_b^∞ functions (Previously, we had $h_o = -\frac{1}{2} Dh$). Write $< f,\mu >$ for $\int f\, d\mu$ and consider the measure-valued equation

$$d < f, \sigma_t > = < Lf, \sigma_t > dt + < Df, \sigma_t > o\, dy_t \qquad (2)$$

where $< f, \sigma . > = f(x)$ i.e. $\sigma_o = \delta_x$. The solution can be expressed as follows: Define x_t by

$$df(x_t) = Zf(x_t).dy_t + Y_o f(x_t)dt + Y_j f(x_t) \circ db_t^j,\ x_o = x,$$

where b^j are Brownian motion independent of y. Then the solution is

$$\sigma_t(f) = E^b[f(x_t)\exp(\int_o^t h_o(x_s)ds + \int_o^t h(x_s) \circ dy_s)].$$

Kunita [17] show that this solution is uniue if coefficients are smooth and bounded. Now the question is whether σ_t has a density.

<u>THEOREM 3.1</u>:

$$< f, \sigma_t > = < B_{y_t} f, \nu_t > \qquad (3)$$

where

$$B_{y_t} f(x) = < f, \mu_t >$$

and μ_t, ν_t *satisfy the equations*

$$d < f, \mu_t > = < Df, \mu_t > \circ dy_t \quad (4)$$

$$< f, \mu_o > \ = f(x)$$

$$d < f, \nu_t > = < B_{y_t} L B_{y_t}^{-1}, \nu_t > dt. \quad (5)$$

<u>PROOF</u>: If $L = o$, then (4) is the same as (2); so the solution of (4) is

$$< f, \mu_t > = f(x_t) \exp (\int_o^t h(x_s) \circ dy_s)$$

where x_t satisfies

$$df(x_t) = Zf(x_t) \circ dy_t.$$

But this has pathwise solution $x_t = \xi(y_t, x)$. The previous definition of B was

$$B_{y_t} f(x) = f(\xi(y_t,x))\exp(\int_o^{y_t} h(\xi(u,x))du).$$

Now,

$$d(\int_o^{y_t} h(\xi(u,x))du) = h(\xi(y_t,x)) \circ dy_t.$$

So (3) holds with B_t defined as before. Now

$$d < B_{y_t} f, \nu_t > = d' < B_{y_t'} f, \nu_t > + < B_{y_t} f, \dot{\nu}_t > dt$$

$$= < B_{y_t} L B_{y_t}^{-1} B_{y_t} f, \nu_t > dt$$

$$+ < B_{y_t} D f, \nu_t > \circ dy_t \quad \text{by (5) and (4).}$$

$$= < B_{y_t} L f, \nu_t > dt + < B_{y_t} D f, \nu_t > \circ dy_t.$$

This verifies (2).

PROPOSITION 3.2: Suppose ν_t has a density function $q_t(z,x)$.[†] Then for $t > o$. σ_t has density

$$\rho_t(V) = q_t(\xi^{-1}(y_t,V),x)\exp\left(\int_o^{y_t} h(\xi^{-1}(s,V))ds\right)$$

$$\times \left|\frac{\partial}{\partial V}\xi^{-1}(y_t,V)\right| \tag{6}$$

where $\left|\frac{d\xi^{-1}}{dV}\right|$ is the Jacobian of the map $V \to \xi^{-1}(y_t,V)$.

PROOF: If ν_t has a density q_t, then

$$< f,\sigma_t > = \int B_{y_t} f(z)q_t(z,x)dz$$

$$= \int f(\xi(y_t,z))\exp\left(\int_o^{y_t} h(\xi(u,z))du\right)q_t(z,x)dz.$$

Changing the variable to $V = \xi(y_t,z)$ gives (6).

THEOREM 3.2: (Bismut [2]) ν_t has C^∞ - density if the Y_i are "smooth" vector fields, i.e., coefficients are bounded with bounded derivatives of all orders, and $Y_1,\ldots,Y_n$ satisfy the "restricted Hörmander condition" H:- Consider vector fields Y_i, $[Y_i, Y_j]$, $[[Y_i, Y_j], Y_k]$ .... At each x, the restrictions of these vector fields to x span $T_x(M)$.

In local coordinates $Y_i = \sum_i b_i(x)\frac{\partial}{\partial x_i}$, ... etc. So the condition says the vectors b etc. span $\mathbb{R}^d$ at each x. Recall, $B_{y_t} L B_{y_t}^{-1} = A_t^* +$ (1st and o^{th} order terms)

$$A^* = \sum_i (\xi_{t*}^{-1} Y_i)^2 + \ldots$$

Now

---

† Here z is the "dummy variable" and x refers to the initial condition in (2), i.e. $\sigma_o = \delta_x$.

$$\xi_{t*}^{-1}[Y_i, Y_j] = [\xi_{t*}^{-1} Y_i, \xi_{t*}^{-1} Y_j] \quad \text{etc.}$$

So if Y_i satisfy the Hörmander condition, then $(\xi_{t*}^{-1} Y_i)$ satisfies it.

Hörmander's own result requires coefficients to be C^∞ in (t,x). Here the coefficients are continuous (but not even C^1) in t. Bismut's version of Malliavin calculus shows that the result still holds with this degree of smoothness in t.

In the filtering problem, the "signal process" involved vector fields $X_1, X_2, \ldots, X_n, X_o$ and $Y = \Delta X$, where Δ is nonsingular if $|\alpha| < 1$. Then $X = \Delta^{-1} Y$. So

$$[\ldots[[X_{i_1}, X_{i_2}], X_{i_3}]\ldots X_{i_k}] = \sum_j c_j [\ldots[[Y_1\delta, Y_2\delta], Y_3\delta] \ldots Y_{i\,n,j}\delta].$$

So if the "X" Lie brackets span $\mathbb{R}^d$ then there must be a collection of "Y" brackets which also span $\mathbb{R}^d$. The Hörmander condition for X with $|\alpha| < 1$ implies the existence of density.

The Case of Vector Observations: Let

$dy^i = h_i(x_t)dt + dW_t^{o,i}$, $W^{o,i}$ are independent Brownian motions. α will now be a matrix,

$$\alpha_{ij}\, t = \langle W^{o,i}, W^{o,j} \rangle .$$

Consider the following cases. (a) Independent signal and noise: Here $\alpha_{ij} = o \;\forall\, i,j$. Then whole theory goes through unchanged.

Then (2) becomes

$$d < f, \mu_t > = \sum_i < h_i f, \mu_t > \circ dy_t^i$$

with solution

$$< f, \mu_t > = \exp\left(\sum_i y_i(t)h(x)\right)f(x)$$

$$= \prod_i \exp\left(y_i(t)h(x)\right)f(x)$$

So this gives pathwise solution as before.

Another Point of View: The Kallianpur-Striebel formula is

$$\sigma_t(f) = E^{(x)}[f(x_t)\exp(\sum_i \int_o^t h_i(x_s)dy_s^i - \frac{1}{2}\sum_i \int_o^t h_i^2(x_s)ds)]$$

$$= E^{(x)}[f(x_t)\prod_i e^{y^i(t)h(x_t)}\exp(\sum_i \int_o^t y^i(s)dh_i(x_s)$$

$$= \frac{1}{2}\sum_i \int_o^t h_i^2(x_s)ds)].$$

(b) The General Case: Here we have no "pathwise" theory (except under very artifical conditions) but the same theory goes through a.s. (Wiener measure). There is no continuous extension to the whole of $C^P[o,T]$. In this case, equation (2) becomes

$$d < f, \mu_t > = \sum_i < D_i f, \mu_t > \circ dy_t^i$$

where

$D_i = Z_i + h_i$ and Z_i is a vector field.

A pathwise solution only exists if D_i's commute, which is very artificial. But, as before, the solution can be expressed as

$$< f,\mu_t > = f(x_t)\exp(\sum_i \int_o^t h_i(x_s)\circ dy_s^i) \qquad (7)$$

where x_t satisfies

$$df(x_t) = \sum_i D_i f(x_t) \circ dy_t^i$$

$$x_o = x.$$

Regard μ as the operator mapping $f \to < f, \mu_t >$. Then "stochastic flow" theory (Elworthy [13], Kunita [17], Bismut [2]) says that if D_i's have smooth coefficients then $x \to x_t(x,\omega)$ is a diffeomorphism a.a. ω, and so the inverse map $x_t^{-1}(x)$ exists. We have to calculate μ_t^{-1}. Generalize (7) slightly to

$$< f,\mu_{s,t} > = f(x_t,t)\exp(\sum_i \int_s^t h_i(x_r,r) \circ dy_r^i)$$

$$df(x_t,t) = \sum_i D_i f(x_t,t) \circ dy_t^i \qquad t \geq s$$

$$x_s = x.$$

PROPOSITION 3.3: (Kunita [17]):

$$\mu_{s,t}^{-1}(f(x)) = f(x_{s,t}^{-1}(x))\exp(-\sum_i \int_s^t h_i(x_{r,t}^{-1}(x)) \circ \hat{d}y_r^i) \quad (8)$$

Where " $\circ\hat{d}$ " means backwards Stratonovich integral. Here define σ - fields $F_{r,t}$ $r \leq t$, by

$$F_{r,t} = \sigma\{ y_u^i - y_v^i,\ r \leq u,v \leq t, i = 1,2,\ldots,d\}.$$

Then $\int_s^t \phi_r \,\hat{d}\, y^i$ is a well defined backward Ito integral if ϕ_r is a backward semimartingale w.r.t. $(F_{r,t})_{r \leq t}$. Then

<u>the Stratonovich integral is defined as usual. If</u> ϕ_r <u>is continuous, then</u>

$$\int \phi_r \circ \hat{d}\, y_r^i = \sum_k \phi\left(\frac{t_k^n + t_{k+1}^n}{2}\right)\left(y^i_{t^n_{k+1}} - y^i_{t^n_k}\right)$$

<u>So</u> $\mu_{s,t}^{-1}(x)$ <u>is well defined by (8). Now verify that</u>

$$\int_s^t h_i(x_{r,t}^{-1}(p)) \circ \hat{d}\, y_r^i \Big|_{p = x_{s,t}(x)} = \int_s^t h_i(x_{s,r}) \circ d\, y_r^i .$$

<u>This checks that</u> $\mu_{s,t}^{-1}(\mu_{s,t}(x)) = x$.

Now all remaining calculations go through as before <u>but only</u> a.s. (Wiener measure).

More general results on existence of densities have been obtained by Bismut and Michel [3].

BIBLIOGRAPHY

[1] R.M. BLUMENTHAL and R.K. GETOOR, Markov processes and Potential theory, Academic press, New York 1968.

[2] J.M. BISMUT, Martingales, The Malliavin calculus and Hypoellipticity under general Hörmander conditions; Z. Wahrscheinlichkeitstheorie 56 (1981) 469-505.

[3] J.M. BISMUT and D. MICHEL, Diffusions conditionelles I, II, J. Functional Analysis 44 (1981), 174-211, 45 (1982) 274-292.

[4] R. BOEL, P. VARAIYA and E. WONG, Martingales on jump processes. I Representation results, SIAM J. Control 13 (1975) 999-1021.

[5] J.M.C. CLARK, The design of robust approximations to the Stochastic differentials equations of nonlinear filtering, Communication Systems and Random Process Theory, ed. J.K. Skwirzynski, NATO Advanced Study Institute Series, Sijthoff and Noordhoff, Alphen aan den Rijn, 1978.

[6] M.H.A. DAVIS, The representation of Martingales of jump processes, SIAM J. Control and Optimization 14 (1976) 623-638.

[7] M.H.A. DAVIS, Piecewise-deterministic Markov Processes, a general class of non-diffusion Stochastic models, to appear in J. Royal Statist. Soc. (B).

[8] M.H.A. DAVIS and S.I. MARCUS, An Introduction to nonlinear filtering, Stochastic Systems: The Mathematics of Filtering and Identification and Applications, ed. M. Hazewinkel and J.C. Willems, 1981.

[9] M.H.A. DAVIS, Pathwise nonlinear filtering, Stochastic Systems, The Mathematics of Filtering and Identification and applications, ed. M. Hazewinkel and J.C. Willems, Reidel (1981) 505-528.

[10] M.H.A. DAVIS, On multiplicative functional transformation arising in nonlinear filtering theory, Z. Wahrscheinlichkeits theorie Verw. Geb. 54, (1980) 125-139.

[11] M.H.A. DAVIS, Linear Estimation and Stochastic Control, Chapman & Hall, London 1977.

[12] C. DELLACHERIE and P.A. MEYER, Probability and Potentials, North-Holland 1978.

[13] K.D. ELWORTHY, Stochastic dynamical systems and their flows, Stochastic Analysis, ed. A. Friedman and M. Pinsky, Academic Press, New York 1978.

[14] R.J. ELLIOTT, Stochastic Calculus and Applications Springer-Verlag 1982.

[15] N. EL KAROVI, Les aspectes probabilistes du contrôle stochastique, in Lecture Notes in Mathematics, 876, Springer-Verlag 1981.

[16] B.V. GNEDENKO and I.I. KOVALENKO, Introduction to the theory of Mass Service [Russian] Moscow, 1966.

[17] H. KUNITA, Densities of a measure-valued process governed by a stochastic partial differential equation, Systems and Control letters 1 (1981) 100-104.

[18] H. KUNITA, Stochastic partial differential equations connected with nonlinear filtering, Nonlinear Filtering and Stochastic Control, Lecture Notes in Mathematics 972, Springer-Verlag, Berlin 1982.

[19] G. KALLIANPUR, Stochastic Filtering theory, Springer-Verlag, 1980.

[20] E. PARDOUX, Equations du filtrage non-linéaire, de la prediction et du lissage, Stochastics 6 (1982) 193-231.

[21] S.R. PLISKA, Controlled jump processes, Stochastic processes and their applications 3 (1979) 259-282.

[22] Z. ROSBERG, P.P. VARAIYA and J.C. WALRAND, Optimal Control of Service in Tandem Queues, IEEE Trans. Automatic Control 27 (1982) 600-610.

[23] H.J. SUSSMAN, On the gap between deterministic and Stochastic ordinary differential equations, The Annals of Probability 6 (1978) 19-41.

[24] D. VERMES, Optimal dynamic control of a useful class of randomly jumping processes, report pp-80-15, International Institute for Applied Systems Analysis, Laxenburg, Australia, 1980.

[25] D. VERMES, Optimal Control of Piecewise-deterministic Process, Preprint, Bolyai Institute, University of Szeged, Hungary 1983 (to appear in Stochastics).

[26] R.B. VINTER and R.M. LEWIS, A necessary and sufficient condition for optimality of Dynamic programming type, Making no apriori assumptions on the controls, SIAM J. Control and Optimization 16 (1978) 571-583.

[27] A.D. WENTZELL, A course in the theory of Stochastic processes, McGraw-Hill Inc. 1981.

[28] E. WONG, Recent progress in Stochastic processes - a survey, IEEE Trans. Inform. Theory, IT-19, 1973, 262-275.

[29] F.A. VAN DER DUYA SCHOUTON, Markov Decision Drift Processes, Mathematical Centre Tract, Amsterdam 1983.

[30] A.A. YUSHKEVICH, Continuous-time Markov decision processes with interventions, Stochastics 9 (1983) 235-274.

[31] R. RISHEL, Dynamic programming and minimum principles for systems with jump Markov disturbances, SIAM J. Control 13 (1975) 338-371.

TATA INSTITUTE OF FUNDAMENTAL RESEARCH

LECTURES ON MATHEMATICS AND PHYSICS

General Editors: S. RAGHAVAN and B.V. SREEKANTAN

MATHEMATICS

1. On the Riemann Zeta-function. By K. Chandrasekharan*
2. On Analytic Number Theory. By H. Rademacher*
3. On Siegel's Modular Functions. By H. Maass*
4. On Complex Analytic Manifolds. By L. Schwartz*
5. On the Algebraic Theory of Fields. By K.G. Ramanathan*
6. On Sheaf Theory. By C.H. Dowker*
7. On Quadratic Forms. By C.L. Siegel*
8. On Semi-group Theory and its Application to Cauchy's Problem in Partial Differential Equations. By K. Yosida*
9. On Modular Correspondences. By. M. Eichler*
10. On Elliptic Partial Differential Equations. By J.L. Lions*
11. On Mixed Problems in Partial Differential Equations and Representations of Semi-groups. By L. Schwartz*
12. On Measure Theory and Probability. By H.R. Pitt*
13. On the Theory of Functions of Several Complex Variables. By B. Malgrange
14. On Lie Groups and Representations of Locally Compact Groups. By F. Bruhat*
15. On Mean Periodic Functions. By J.P. Kahane*
16. On Approximation by Polynomials. By J.C. Burkill*
17. On Meromorphic Functions. By W.K. Hayman*
18. On the Theory of Algebraic Functions of One Variable. By M. Deuring*

---

*Not available.

19. On Potential Theory. By M. Brelot*

20. On Fibre Bundles and Differential Geometry. By J.L. Koszul*

21. On Topics in the Theory of Infinite Groups. By B.H. Neumann*

22. On Topics in Mean Periodic Functions and the Two-Radius Theorem. By J. Delsarte*

23. On Advanced Analytic Number Theory. By C.L. Siegel+

24. On Stochastic Processes. By K. Ito

25. On Exterior Differential Systems. By M. Kuranishi*

26. On Some Fixed Point Theorems of Functional Analysis. By F.F. Bonsall*

27. On Some Aspects of p-adic Analysis. By F. Bruhat*

28. On Riemann Matrices. By C.L. Siegel*

29. On Modular Functions of One Complex Variable. By H. Maass

30. On Unique Factorization Domains. By P. Samuel*

31. On The Fourteenth Problem of Hilbert. By M. Nagata

32. On Groups of Transformations. By J.L. Koszul*

33. On Geodesics in Riemannian Geometry. By M. Berger*

34. On Topics in Analysis. By Raghavan Narasimhan*

35. On Cauchy Problem. By S. Mizohata

36. On Old and New Results on Algebraic Curves. By P. Samuel*

37. On Minimal Models and Birational Transformations of Two Dimensional Schemes. By I.R. Shafarevich*

38. On Stratification of Complex Analytic Sets. By M.-H. Schwartz

39. On Levi Convexity of Complex Manifolds and Cohomology Vanishing Theorems. By E. Vesentini

40. On an Introduction to Grothendieck's Theory of the Fundamental Group. By J.P. Murre*

41. On Topics in Algebraic K-Theory. By H. Bass*

42. On the Singularities of the Three-Body Problem. By C.L. Siegel

---

+ Available as volume 9, T.I.F.R. Studies in Mathematics

43. On Polyhedral Topology. By J.R. Stallings*

44. On Introduction to Algebraic Topology. By G. de Rham

45. On Quadratic Jordan Algebras. By N. Jacobson*

46. On the Theorem of Browder and Novikov and Siebenmann's Thesis. By M.A. Kervaire

47. On Galois Cohomology of Classical Groups. By M. Kneser*

48. On Discrete Subgroups of Lie Groups. By G.D. Mostow

49. On the Finite Element Method. By Ph. Ciarlet

50. On Disintegration of Measures. By L. Schwartz

51. On Introduction to Moduli Problems and Orbit Spaces. By P.E. Newstead

52. On Numerical Methods for Time Dependent Equations - Application to Fluid Flow Problems. By P. Lascaux

53. On Optimisation - Theory and Algorithms. By J. Cea

54. On Deformations of Singularities. By M. Artin*

55. On Sieve Methods. By H.-E. Richert

56. On Irregularities of Distribution. By W.M. Schmidt

57. On Expansion Techniques in Algebraic Geometry. By S.S. Abhyankar

58. On Torus Embeddings and Applications. By T. Oda

59. On Forms of Higher Degree. By J.-I. Igusa

60. On Curves on Rational and Unirational Surfaces. By M. Miyanishi

61. On Wave Propagation. By G.B. Whitham

62. On Equations Defining Space Curves. By L. Szpiro

63. On Topics in Finite Element Solution of Elliptic Problems. By B. Mercier

64. On Diffusion Problems and Partial Differential Equations. By S.R.S. Varadhan

65. On Numerical Methods for Non-linear Variational Problems. By R. Glowinski

66. On Representations of Complex Semi-simple Lie Groups.
By T.J. Enright

67. On Nonlinear Waves and Shocks. By C.S. Morawetz

68. On Topics in Stochastic Differential Equations.
By D.W. Stroock

69. On Moduli of Curves. By D. Gieseker

70. On Partial Differential Equations. By G.B. Folland

71. On Three-dimensional Elasticity. By P.G. Ciarlet

72. On Sieve Methods and Prime Number Theory. By Y. Motohashi.

73. On Stochastic Differential Equations and Malliavin Calculus.
By S. Watanabe

74. On Results on Bezout's Theorem. By W. Vogel

75. On Stochastic Control and Nonlinear Filtering. By M.H.A. Davis